THE PEARLS OF COMMUNICATION, HISTORY TAKING, AND PHYSICAL EXAMINATION

THE PEARLS OF COMMUNICATION, HISTORY TAKING, AND PHYSICAL EXAMINATION

THE ROAD TO PASSING CLINICAL EXAMINATIONS

MOHAMED-ELBAGIR KHALAFALLA AHMED

MBBS(University of Khartoum, Sudan) MD MRCP(UK)
FRCP(London) FACP(USA) FAATM(USA)

Professor of Medicine, College of Medicine
King Khalid University, Abha, Saudi Arabia

Library of Congress Control Number: 2017901779
ISBN: Hardcover   978-1-5245-2206-3
                     Softcover   978-1-5245-2205-6
                     eBook   978-1-5245-2229-2

Print information available on the last page.

Rev. date: 02/13/2017

To order additional copies of this book, contact:
Xlibris
1-800-455-039
www.Xlibris.com.au
Orders@Xlibris.com.au
732084

TABLE OF CONTENTS

Chapter 3: The Art of Physical Examination

DEDICATION

To all those who have strived to teach us the art of clinical medicine.

To the souls of the late Professor Daoud Mustafa Khlaid and the late Professor Siddig Ahmed Ismaeel and all who taught us over the years from whom I learned the highest quality bedside medicine and manners.

To our medical students everywhere and to our young colleagues and trainees who need such pearls to help them practice fine medicine and pass their examinations.

This book is dedicated.

Mohamed Elbagir Khalafalla Ahmed MBBS MD MRCP FRCP FACP FAATM

Professor of medicine, January 2017

FOREWORD

Optimal patient care necessitates that doctors should be competent in key educational areas, mainly communication skills, history taking, and performing through physical examination. Acquiring such fundamental skills needs perfect teaching and reading good books. Actually, there are few books that deal with these fundamental issues in detail; hence, I have compiled those in this text using relevant clinical information, practical examples, and examination tips. These will be a great help to both medical students and postgraduate candidates.

In general, doctors taking a postgraduate examination should have fulfilled the following:

(1) Completed an adequate period of training after graduation (at least four years).
(2) Have been well-supervised during the period of training according to a well-structured, postgraduate training programme.
(3) Have gained adequate clinical experience and competency so that they are able to elicit comprehensive medical history and perform thorough physical examination in a skillful manner.
(4) Are able to deliver emergency medical care, and most importantly, comprehensive patients management.

(5) To fulfill that, they should be able to effectively communicate with patients, hospital staff, and colleagues using the well-recognised communication skills and ethical approaches.

To achieve that, medical students and doctors in training need a guide and help. Besides formal teaching and major textbooks, this modest endeavour will greatly help students and doctors to prepare themselves for postgraduate exams (also undergraduate exams) in internal medicine, and above all, will also improve patient care during clinical practice.

Thus, this book offers practical help in simple format, which is written by an experienced academic and teacher with over thirty-five years of experience in the field of internal medicine teaching and examinations worldwide.

ACKNOWLEDGEMENT

I sincerely thank the following for their kind permission to quote some of the contents used in this book. I would also like to extend my thanks to all those who provided me with excellent material which I included in this book which is intended to improve the skills of our students & doctors.

1. Travaline, John M.
2. Tom Ferris. site; www.bppuniversity.ac.uk/medicalseries
3. graham-graham@reikiworld.net
4. Matt Green, Teresa Parrott, Graham Crook.
5. Denise M. Dudzinski, PhD MTS

   University of Washington. dudzine@u.washington.edu.
6. bhinfo@uw.edu. www.depts.washington.edu/bhdept
7. Brusch JL. Infective Endocarditis. P 79, 82-83. *Medscape Drugs & Diseases.*

   http://emedicine.medscape.com/article/216650-overview. Accessed Apr 6, 2016.

8. Shaffrey, Chris.
9. Mac master university, Canada:

CHAPTER 1

Communication Skills and Ethics: Gaining Competency

Overview

'The patient will never care how much you know until they know how much you care.'

Effective doctor-patient communication is a central clinical function in building a therapeutic doctor-patient relationship, which is the heart and art of medicine. This is important in the delivery of high-quality health care. Much patient dissatisfaction and many complaints are due to breakdown in the doctor-patient relationship.

Patient-physician communication is an integral part of clinical practise. When done well, such communication produces a therapeutic effect for the patient, as has been validated in controlled studies. When doctors use communication skills effectively, both they and their patients benefit.

- Doctors identify their patients' problems more accurately
- Their patients are more satisfied with their care and can better understand their problems, investigations, and treatment options.
- Patients are more likely to adhere to treatment, and to follow advice on behaviour change.

- Patients' distress and their vulnerability to anxiety and depression are lessened.
- Doctors' own well-being is improved
- Doctors are less likely to be sued and go to court

From obtaining the patient's medical history to conveying a treatment plan, the physician's relationship with his patient is built on effective communication. In these encounters, both verbal and non-verbal forms of communication constitute this essential feature of medical practise. Although much of the communication in these interactions necessarily involves informationsharing about diagnosis and therapy options, most physicians will recognise that these encounters also involve the patient's search for a therapeutic relationship.

Physicians should become competent in five key communication skills:

1. Listening effectively
2. Eliciting information using effective questioning skills (e.g., history taking)
3. Providing information using effective explanatory skills
4. Counseling and educating patients
5. Making informed decisions based on patient information and preference
6. Breaking bad news

How Effective Communication Makes a Good Doctor

Listening Skills

A doctor who patiently listens, and who is much interested to address the concerns of the patients, will be highly rated by patients. This skill will make the patients trustand be closer to the doctor. Below are summaries of some of the expressions raised by patients about their doctors.

- My doctor attentively listens and does not ignore my feelings.

- The doctor I see is very encouragingand allows us to ask questions and request explanations.
- I felt that the doctor is very caring and gets to the bottom of my problem.
- We are lucky to have such a patient doctor who allows us to explain our problem without interruption.

The Partnership

It is a really a true quality when a doctor creates a sincere partnership with the patient by using listening skills, attending topatient's concerns, and respecting thevalues of patients. This will facilitate the patients to openly discuss their concerns.

Also, allowing the patient to share opinions about the management of his disease is a top communication skill referred to as patient autonomy.

Other Qualities of a Good Communicator:

Has respect for her patient. Good doctors understand that a sick or injured patient is highly vulnerable. Being respectful goes a long way towards helping that patient.

Has the ability to share information in terms her patients can understand.

Doesn't interrupt or stereotype her patients. It's easy for all of us to interrupt when we know time is short or we are in a hurry, but a practitioner who is a good communicator knows that if it can't be done right, to begin with, it will need to be done over.

Has the ability to effectively manage patients' expectations. By helping the patient understand what the next steps will be, and what the possible outcomes and their ramifications might be.

A poor communicator will criticize, belittle, or disrespect the patient by saying:

'You worry too much.'
'You got upset over nothing.'
'You've got to try to cooperate.'
'I'd like you to be more responsible.'(Implies patient is irresponsible)
'You're just too lazy.'

Difficult Patients Situations in Medical Encounters

Silent patient
Rambling or talkative patient
Vague patient
Angry patient.
Depressed or sad patient
Denial

- Anxious patient
- Somatization

Types of Communication

A. Verbal: talking, using words. This is the most common.
B. Non-Verbal: very important. This includes:

   a. Facial expression: smile, using lips, head nodding
   b. Keeping eye contact but not gaze
   c. Posture: bend to impress the other side
   d. Distance: average, not too farand not too close
   e. Touch: to comfort and to ease tension

Types of Questions Used in Communication:

1. **Open-ended:** Goodquestions that are answered by a statement, not answered by yes or no. Example:'Tell me about the character of your pain.'

2. **Closed questions**: Not recommended, use minimally. 'Do you have pain?Yes or No.'

3. **Leading:** You want the patient to answer what you have in mind. Do not use 'your pain is stabbing, yes?'

4. **Explanatory:** Good. You request more information. 'You say you are not feeling well. Tell me more about this.'

5. **Compound or multiple:** Not recommended. 'Do you have pain, nausea, vomiting, or indigestion?'

Dealing with Emotions

This is a key communication skill. Doctors should sense and respond appropriately to patient concerns and emotions. This is achieved by:

Empathy: Sharing the patient feeling and providing support. It is caring, understanding, and supporting.

- Use empathy to communicate understanding and appreciation of the patient's feelings or predicament, and overtly acknowledges patient's views and feelings
- Provides support: expresses concern, understanding, and willingness to help; acknowledges coping efforts and appropriate self-care; offers partnerships
- Deals sensitively with embarrassing and disturbing topics and physical pain including when associated with physical examination

Barriers to Effective Communication

It is not always easy to communicate with patients for a number of reasons:

- Speech problems
- Spoken foreign language
- Time constraints on physician or patient
- Unavailability of physician or patient to meet face to face
- Altered mental state
- Psychologic or emotional distress
- Gender differences
- Racial or cultural differences

Communication and Patient Outcomes

Research on the effect of communication skills on patients has focused on three areas: patient satisfaction, patient concordance, and physiological outcomes. These three are, of course, closely connected. For example, the patient who has a clear understanding of the nature of their hypertension or diabetes and how to manage it, both through lifestyle and pharmacological means, is more likely to follow the recommended management plan accurately, resulting in better control. It has also been suggested that an effective doctor-patient relationship can be therapeutic in itself.

Patient Satisfaction

Patient satisfaction has been linked to a partnership relationship between doctor and patient where there is a human connection, and the doctor demonstrates empathy, a non-judgemental attitude, and warm non-verbal communication. A patient-centred approach, in which the patient's perspective, including beliefs, fears, and expectations, is actively integrated into the consultation, and is also associated with increased satisfaction. A patient-centred approach leads to a greater understanding of the patient's reasons for seeking help.

Researchers have identified that the single most powerful predictor of patient satisfaction is the amount of jargon-free information the patient receives on diagnosis, causes, prognosis, and possible treatments. Appropriately delivered and timely information helps patients to cope with the uncertainty of illness, and participate in the management of their illness.

Clinical competence also, not surprisingly, leads to patient satisfaction, although some studies have suggested patients are unable to objectively assess technical competence except in glaring examples, and tend to infer it from the doctor's interpersonal skills. Paling suggests that patient trust is a function of perceived competence and caring.

Adherence

Non-concordance rates with prescribed medication have been estimated to be between 30–60per cent in a range of studies with the most significant determinants of concordance being the patients' understanding of the illness, the rationale of the treatment, their understanding of the drug regimen, and their relationship with and trust in the doctor. Eliciting the patient's knowledge, beliefs, and concerns about their illness also increases adherence to treatment regimens.

Symptom Relief and Physiological Outcomes

A number of different studies have demonstrated an association between doctor-patient communication and physiological measures such as:

- Resolution of chronic headache
- Post-surgical recovery and decreased length of hospitalisation
- Post-myocardial analgesia

- Blood pressure and blood sugar measurements
- Improved symptoms and function of patients with rheumatoid arthritis and peptic ulcer disease
- Perceived health status and daily functioning
- Duration of sore throat
- The possibility of improved survival rates from breast cancer

Important Steps of Effective Communnication

- Intiating the interview by making rapport with the patientThis is done by:

  (a) greeting the patient;
  (b) introducing yourself: mention your name and grade;
  (c) secure patient comfort, privacy, and welfare;
  (d) calling the patient by his/her pfeferred name.

- Using appropriate and clear language by avoiding the use of medical and scientific terms that are difficult for the patient to understand. Examples: Do not say the cancer has metastasized to the bone;better say 'I am sorry that your tumour has spread to the bone.'
- Use body language that is non-verbal:

  To be interested in the patient, maintain eye contact and do not be distracted by looking at the notes or the door. Frequently use facial expressions including smiling and head nodding. This will leave a positive impression on the patient.

- Discuss and agree on what you will be discussing with the patient (the agenda).

  Mention to the patient what you will be discussing and ask him/her if they have anything to add.

Agreement:

> Always summarise and agree with patient on what has been discussed. Make sure the patients accept the information, whether it is a diagnosis or a management plan. This will lead to improved patient complaince.

- Be open-minded and supportive:

  Patients will feel comfortable and trust the doctor if the latter has a flexible and supportive attitude. Listen and listen so that you can grasp the feelings and concerns of the patient.

- Give informative account of the patient's condition:

  Avoid talking too much about details of diagnostic tools and drug therapy. Be practical and provide the necessary information that will make the patient reassure and well informed. Patients like the doctor who tells them about practical issues when it comes to drug therapy and its side effects. Explain to them whatthey do if anything go wrong. This will reduce stress and improves psychological status and hence improves the quality of life.

- Be honest, admit, and apologise when mistakes occur:

  By doing so, patients will be in a better way not to sue you or your foundation.

  The comments cause of litigation is ineffective communication. Thus, good communication and open discussion with apology will lessen the anger and concerns of the patients, and will decrease the chances of litigation.

- Use empathy and listening skills:

  Again, I emphasise the importance of listening to the patient. This will allow the patient to express himself and feel that you are ready to help. Empathy means sharing the patient's concerns and feelings and providing full support and showing that to him/her. Words like 'I feels just as you do, and I fully share you these feeling and will work together to give you maximum support,'this will improve your relation with the patient and improve therapeutic ties.

- Be attentive and mindful:

  Observe what the patent says or feel, as well as observe and periodically evaluate your own performance. Check whether you have provided the right information in the appropriate manner or not. Check the words and the style you used and ammend it if necessary.

- Dealing with emotions:

  This is a vital and a very important issue. During the conveying of information, especially when breaking bad news, sense and observe the emotions of the patients. This can be anxiety, anger, depression, or emotional breakdown. Thus, appropriately deal with these emtions by showing empathy and support. Avoid giving false or premature reassurances and attending to the physical illness alone, as these will further worsen the situation.

Protocols of Communication

For General Communication, the CLASS Protocol

About 'CLASS'

'C' for Context

The CLASS and SPIKES protocols both begin with attention to the 'context', or 'setting' in which the interaction will take place. Here are some valuable pointers for getting it right.

Quiet room, sit the person and make him/her comfortable. Keep a reasonable distance, invite others to attend if agreed, and maintain privacy.

L for Listening Skills

Once the setting is right, other techniques come into play. The first has to do with listening skills—the *L* of the CLASS acronym. The GMC in Britain stresses this skill by mentioning, 'Doctors must listen to patients, take account of their views, and respond honestly to their questions.'

A for Acknowledging

The *A* of CLASS stands for acknowledging the patient's emotions. The technique for doing this—the empathic response—is outlined here. This is a critical skill.

Definition: Empathy is the ability to see the world as seen by the other person, to share and understand another person's feelings, needs, concerns, and/or emotional state.

Empathy is a selfless act that enables us to learn more about people and relationships with people. It is a desirable skill beneficial to ourselves, others, and society. Phrases such as 'being in your shoes' and 'soulmates' imply empathy. Empathy has even been likened to a spiritual or religious state of connection with another person or group of people.

There is an important distinction between empathy and sympathy.

We offer our sympathy when we imagine how a situation or event was difficult or traumatic to another person. We may use phases like 'I am very sorry to hear that' or 'If there is anything I can do to help. . .' and we feel pity or sorry for the other person. This is how many people would react to the famine example above. There is nothing wrong with sympathy, and it can help to offer some help.

11

To empathise is to feel how others feel, to see the world as they do, or to be 'in their shoes'.

Example: 'I really feel same as you do, and I share yourfeeling. Also, we will do our best to support you.'

S for Strategy

The first *S* of CLASS stands for strategy, the forming of a plan.

S = Strategy: You have to agree with the other side on a roadmap or plan for managing his/her problems.

S for Summary

The second *S* of CLASS stands for summary. This is very important and very often left undone in clinical interviews and interactions.

S = Summary

Key Points

1. Prepare the scene
2. Assess what the patient already knows

    Before providing information, find out what a patient already knows about his or her condition.

3. Assess what the patient wants to know

    Not all patients with the same diagnosis want the same level of detail in the information offered about their condition or treatment.

4. Be empathic

   Empathy is a basic skill physicians should develop to help them recognise the indirectly expressed emotions of their patients.

5. Slow down, reasonable pace

   Physicians who provide information in a slow and deliberate fashion allow the time needed for patients to comprehend the new information.

6. Make it simple and avoid jargon (medical terms)
7. Be honest and tell the truth
8. Be hopeful

   Although the need for truthtelling remains primary, the therapeutic value of conveying hope in situations that may appear hopeless should not be underestimated. Particularly in the context of terminal illness and endoflife care, hope should not be discouraged.

9. Watch the patient's body and face

   Much of what is conveyed between a physician and patient in a clinical encounter occurs through non-verbal communication. For both physician and patient, images of body language and facial expressions will likely be remembered longer after the encounter than any memory of spoken words.

10. Be prepared for a reaction

    Patients vary not only in their willingness and ability to absorb information, but also in their reactions to physician communications.

Bad Communication, Avoid:

1. Using technical language or jargon when communicating with patients
2. Ignoring appropriate concern for the patient's problems
3. Not applying listening skills
4. Failure to verify that the patient has understood the information
5. Not applying empathy
6. Being apathetic or not interested in the patient
7. Using poor language and not applying body language

Breaking Bad News

- Is an important communication skill
- Is a complex communication task which includes:

  - Responding to patients' emotional reactions
  - Involving the patient in decision-making
  - Dealing with the stress created
  - Involvement of multiple family members
  - How to give hope when situation is bleak

The imparting of bad news is a key role in a doctor's job. It is a communication skill that, like all others, must be learnt and honed. Medical schools have focused intently on communication in recent years and post–graduate college exams will involve a communication element. This is evidence of the recognition, albeit perhaps delayed, that such skills are central to our work.

Though it may be awkward, the only way to learn is to watch those who are more experienced. Like any other skill in our profession, you must observe and note which you consider good and that you would hope to imitate.

You have to be in the room when bad news is broken, and only if the family or patient have no hint of an objection, but you must be there.

No one should ever be alone when ill tidings are in the air. This includes you. Patients should have the opportunity to have someone with them, and so should you.

When you state that the person has died, do not rush to say any more than 'I'm very sorry'. Even if the news was expected, there must be some time to allow it to sink in, and nothing you say in those first few moments will be taken in. Do not be tempted to go into great detail unless the family request it and question you.

A further meeting later that day or the next may be the time to discuss matters in depth. I have seen families in bleary bewilderment, while someone tries to describe in detail why the emergency surgery was unsuccessful. Anatomical terms, operative equipment, and procedural names—the whole works. This is rarely appropriate.

How to Do It?

SPIKES Protocol

Practical situation:

MrH. A., aged seventy-three, has had a prostate biopsy following a screening PSA. The results showed that he has a moderately differentiated adenocarcinoma (Gleason Score 7). He has not been given much information prior to this interview. In this vignette, the physician will disclose the biopsy results.

Communication Issues

Disclosing a cancer diagnosis is a common 'breaking bad news' situation. It is often not easy, so it helps to know where to start, and how to proceed in giving information and dealing with the patient's responses to the news.

How to Break Bad News?

In the interview, you will see the physician follow the steps of SPIKES. Having attended to the setting (S), he asks for the patient's perception of his situation (P), gets an invitation to proceed before disclosing news(I), only then giving the information or knowledge (K). Note that the physician remembers at all times to deal emphatically with any emotion that arises (E), and closes by offering a treatment strategy and summary or plan for going forward (S).

The SPIKES Protocol

In the interview, you should follow the steps of SPIKES:

S: Settting the scene: Introduce self, explain purpose, confirm the name of the patient/person, select a quiet room(privacy), and ask other members of the team or patient relative to attend.

P: Ask for the patient's perception of his situation. What the patient already knows or he has been told about his condition.

I: Get an invitation to proceed before disclosing news. Invite the patient if he wants to know all the details or just the summary.

K: This is the main discussion, laying the knowledge of the condition. Give the information in small steps and pause to sense the reaction of the patient.

Step 4: K – giving *knowledge* and information to the patient.

Warning shot: say 'I am afraid I have bad news for you...'

- Use simple language, no jargon(do not use medical terminology)
- Vocabulary and comprehension of patient
- Small chunks, avoid detail unless requested
- Silence, pause, allow information to sink in. Do not interrupt.
- Wait for response before continuing

- Check understanding
- Check impact

E: Note that the physician remembers at all times to deal emphatically with any emotion that arises during discussion of the condition. Empathy means you tell the patient that you shre his feelings and will do all you can to support him/her. Deal with any emotions like crying, feeling depressed, etc.

Step 5: E – addressing the patient's *emotions* with empathic responses

- Shock, isolation, grief
- Silence, disbelief, crying, denial, anger
- Observe patient's responses and identify emotions
- Offer empathic responses

S: Then close the session by offering a treatment strategy and summary and a manegement plan that includes follow-up.

Step 6: S – *strategy* and *summary*

- Are they ready?
- Involve the patient in the decision-making
- Check understanding
- Clarify patient's goals
- Summarise
- Contract for future

Summary Points

1.	Assess the patient's understanding first: what the patient already knows, is thinking, or has been told.
2.	Gauge how much the patient wishes to know
3.	Give warning first that difficult information are coming (e.g., 'I'm afraid we have some work to do', 'I'm afraid it looks more serious than we had hoped.')

4. Give basic information, simply and honestly; repeat important points

5. Relate your explanation to the patient's framework

6. Do not give too much information too early; don't pussyfoot, but do not overwhelm

7. Give information in small 'chunks'; categorize information-giving

8. Watch the pace, check repeatedly for understanding and feelings as you proceed

9. Use language carefully with regard given to the patient's intelligence, reactions, emotions, and avoid jargon.

Being Sensitive to the Patient

- Read the non-verbal clues: face/body language, silences, tears
- Allow for 'shut down' (when patient turns off and stops listening), and then give time and space; allow possible denial.
- Keep pausing to give patient opportunity to ask questions.
- Gauge patient's need for further information as you go, and give more information as requested(i.e., listen to the patient's wishes as patients vary greatly in their needs).
- Encourage expression of feelings, give early permission for them to be expressed(i.e., 'How does that news leave you feeling', 'I'm sorry that was difficult for you', 'You seem upset by that')
- Respond to patient's feelings and predicament with acceptance, empathy, and concern
- Check patient's previous knowledge about information given.
- Specifically elicit all the patient's concerns.
- Check understanding of information given ('Would you like to run through what are you going to tell your wife?')
- Be aware of unshared meanings (i.e., what cancer means for the patient compared withwhat it means for the physician).
- Do not be afraid to show emotion or distress.

Planning and Support

- Having identified all the patient's specific concerns, offer specific help by breaking down overwhelming feelings into manageable concerns, prioritising and distinguishing the fixable from the unfixable.
- Identify a plan for what is to happen next.
- Give a broad time frame for what may lie ahead.
- Give hope tempered with realism ('preparing for the worst and hoping for the best').
- Ally yourself with the patient:'We can work on this together...between us.'(i.e., co-partnership with the patient/advocate of the patient)
- Emphasise the quality of life.
- Safety net

Follow-Up and Closing

- Summarise and check with patient.
- Don't rush the patient to treatment.
- Set-up early further appointment, offer telephone calls, etc.
- Identify support systems; involve relatives and friends.
- Offer to see/tell spouse or others.
- Make written materials available.

Remember: doctor's anxiety - regiving information, previous experience, failure to cure, or help

Dealing with Special Situations

What if the patient starts to cry while I am talking?

In general, it is better simply to wait for the person to stop crying. Express empathy and give tissue paper. If it seems appropriate, you can acknowledge it (*'Let's just take a break now until you're ready to start again'*), but do not assume you know the reason for the tears (you may want to explore the reasons now or later).

Most patients are somewhat embarrassed if they begin to cry and will not continue for long. It is nice to offer tissue if they are readily available (something to plan ahead), but try not to act as if tears are an emergency that must be stopped, and don't run out of the room—you want to show that you're willing to deal with anything that comes up.

Examples:

> Situation 1:
>
> JM is a 60-year-old man who just had a needle biopsy of the pancreas showing adenocarcinoma. You run into his brother in the hall, and he begs you not to tell his brother, JM, because the knowledge would kill him even faster. A family conference to discuss the prognosis is already scheduled for later that afternoon.
>
> Would you follow the wish of his brother, or respect patient autonomy?

Discussion:

It is common for family members to want to protect their loved ones from bad news, but this is not always what the patient himself would want. It would be reasonable to tell Jose's brother that withholding information can be very bad because it creates a climate of dishonesty between the patient and family and medical caregivers. Also, the only way for Jose to have a voice in the decision-making is for him to understand the medical situation. Ask Jose how he wants to handle the information in front of the rest of the family, and allow for some family discussion time for this matter.

In some cultures, it is considered dangerous to talk about prognoses and to name illnesses. If you suspect a cultural issue, it is better to find someone who knows how to handle the issue in a culturally sensitive way than to assume that you should simply refrain from providing medical information. For many invasive medical interventions, which require a patient to critically weigh burdens and benefits, a patient will need to have some direct knowledge of their disease in western terms in order to consider treatment options.

> Situation 2:
>
> Consider this: You are a 28-year-old female resident in a rotation in an HIV service. FS is a 32-year-old woman with advanced HIV who contracted HIV from her her boyfriend who has been on vacation in an African country. She came to the clinic to discuss the problem.

How would you proceed?

- Tell her the result directly to avoid wasting time.
- Use the SPIKES protocol to convey the result.
- Givea typed copy of the result when you meet her.
- Once you meet her, tell her that she has a serious disease andbook anoutpatient appointment with consultant.

Obviously, you have to follow a well-known protocol, which is the ABCD or the SPIKES.

Although the protocol for breaking bad news is helpful, it doesn't cover everything. Breaking the bad news bluntly will lead to many problems like denial, disbelief, shock, or displacement. Following the SPIKES protocol needs good training and practise.

The Difficult/Angry Patient:

'Difficult patients' can be seen as a problem to be tolerated or terminated from practise however the difficulty is in the *relationship*, not simply the patient, and there are techniques and strategies to help clinicians improve that relationship and retain its therapeutic nature.

What is a Difficult Patient-Clinician Relationship?

This is a common problem encountered in medical practise. A difficult patient-clinician relationship, occurring in approximately 15 per cent of adult patient situations (Krebs et al., 2006), arises when physicians encounter patients with complex, often chronic, medical issues (such as chronic pain and/or mental illness) that are influenced or exacerbated by social factors (such as poverty, abusive relationships, addiction).

Dealing with such patients needs training, experience, and a professional attitude. A strained situation like this may lead to serious consequences if not handled with care. If you clash or react angrily with the patient, the problem may escalate.

Previous experience with similar patients, along with the social and economic disparities between the physician and patient, may make the physician uncomfortable. This may lead the physician to be guarded or distant, which the patient may interpret as distrust.

The physician may become frustrated or angry because his advice is not followed, and because the diagnosis or treatment is unclear or ineffective, or because the patient is rude, seemingly ungrateful, or transgresses boundaries in the clinician-patient relationship (e.g., comes to the clinic when she does not have an appointment). Clinicians may become angry and avoid, or sometimes, 'punish' the patient.

Patients are labelled 'difficult' based on the feelings they invoke in clinicians, such as anger, frustration, anxiety, dread, and guilt. Patients who, for medical or non-medical reasons, appear ungrateful or frivolously utilise medical care are most likely to be described as difficult.

Usually, such patient labelled as difficult may have a reason to be difficult or angry. They may continue to seek medical attention, but wiyout complying with the advice they are given. Often, such patients may have multiple medical complaints, psychiatric conditions (helplessness, depression, anxiety, self-loathing), abrasive personality traits (expressing rage, inflexibility), personality disorders, addictions, and multiple physical symptoms of unknown or ambiguous etiology.

In particular, be careful when dealing with a drug addict or a psychiatric patient. They often make requests that clinicians think are inappropriate, such as requests for additional pain medicine, increased phone contact or clinic appointments, etc.

How to Deal with Difficult Patients

Strategies for Maintaining a Therapeutic Relationship (Krebs et al., 2006; Wasan et al., 2005; Elder et al., 2006; Hass et al., 2005)

1. Do not confront the patient. Be compassionate and empathic. Keep in mind that most patients whom you find frustrating to deal with have experienced significant adversity in their lives.
2. Listen patiently and pick up the reason for the patient to be angry.
3. Acknowledge and address underlying mental health issues early in the relationship.
4. Prioritise the patient's immediate concerns, and elicit the patient's expectations of the visit and their relationship with you.
5. Set clear expectations, ground rules, and boundaries, and stick to them. Have regular visits, which helps convey confidence that the patient can deal with transient flare-ups without an emergency visit.
6. Be aware that strong, negative emotions directed at you are often misplaced. The patient may be imposing feelings and attitudes on

to you that they have had towards other doctors, friends, family members in the past. This is known as transference. Acknowledge the patient's feelings and set behavioural expectations.

7. Beaware of your own emotional reactions, and attempt to remove yourself so you can objectively reflect on the situation. Involve colleagues. Vent your feelings or debrief confidentially with a trusted colleague so that your negative emotions are kept at bay during patient encounters.

8. Recognise your own biases. For example, patients with addictions genuinely need medical care, but the behaviours associated with addiction are vexing for health care providers. These patients are often both vulnerable and manipulative. Be sure that you are attentive to their vulnerability rather than focusing exclusively on their manipulative behaviours.

9. Avoid being very directive with these patients. A tentative style tends to work better. Remember that you provide something many of these patients do not have—a steady relationship with someone who genuinely wants to help them. This, in itself, can improve the patient's health even in the absence of medical treatment.

10. Prepare for these visits. Keep in mind your goals of care and make a strategy for the encounter before it occurs.

11. Give such difficult patients the maximum care, so as not to miss any diagnosis or do harm.

Practical Situations of Difficult Patients

Example 1

Mr A. M. is a 64-year-old man with multiple complex medical problems including uncontrolled diabetes, untreated depression, coronary artery disease (status post-myocardial infarction with multiple stents), painful peripheral vascular disease, hypertension, hyperlipidemia, chronic hepatitis C, spinal stenosis, and ongoing two pack per day tobacco use.

He is on a long list of sixteen medications and insulin, though does not take any of his medications with regularity. Mr D returns to clinic, and all of the above medical problems are not well controlled. He has not been taking medication or following up with specialty care as advised. He continues to smoke, and has not improved his extremely sedentary activity level. He requests that something more be done, implying that your care for him is not optimal.

How would you deal with this patient?

Clearly, this patient has a real problem. He has multiple medical problems and using polypharmacy. This patient has a real reason to be frustrated and confused. Deal with this patient with extreme empathy and sympathy.

Suggestions:

1. Be aware of your own feelings. It is frustrating to devote time and compassion to patients when they do not take your advice.
2. Still, you provide essential service to the patient by listening to his concerns and providing an opportunity for care.
3. Prepare for the visit by acknowledging your frustrations before seeing the patient, and strategise about how you will make the most of his visit. Set your own goal for the visit.
4. Also, ask him why he is coming to see you. If it is unrealistic to expect Mr D to comply with all his medical therapies, if he feels respected and heard, he will be more likely to heed your advice.
5. Consider asking which of his medical problems is bothering him most, and negotiate with him on ways that he can work towards mutually agreed upon goals. Explore causes of his non-compliance, and problem-solve together towards a more coherent medical plan.
6. Emphasise that the two of you work together to improve his health, and establish one or two goals for his next visit (e.g., better diabetes control). Plan to maintain a patient and respectful attitude even though you might actually be feeling frustrated.

7. Ask help from other providers (pharmacist, nurse, counselor, or social worker). This will make a team rather than a single doctor approach.

Example 2

Mr M is a 47-year-old man who comes into clinic to follow-up on multiple uncontrolled medical problems including diabetes, hypertension, obesity, depression, and sleep apnea. He is unemployed and is currently homeless. On further discussion, the patient shares that he lost his last job due to recurrent conflict with a co-worker. He shares his frustration that 'everyone is out for me because they arebad people'. On several occasions, he was aggressive and violent to the reception desk personnel if he did not get the appointment he needed. He wonders why he is always getting kicked out of places and feels he has cause and a 'right' to be angry. Though he seeks ongoing care, he does not have insight into how his anger is perceived by others.

What is the strategy for dealing with such patients?

1. Prepare for your visit with Mr. B by 'venting' and strategising with trusted colleagues.
2. Try calming the patient by saying, 'I understand that you are angry, and I would like to spend some time talking with you about that, but I will immediately end our appointment if there is any threatening speech or behaviour.'
3. Validate some of his feelings by agreeing any form of discrimination is wrong, and he has every reason to be angry, but that learning different ways to express his anger may help minimize the kinds of social interactions that he dislikes.
4. Discuss treatment modalities that may be beneficial to him in this regard. If possible, schedule his visit before your lunch or a break, so you attend to yourself after the visit.
5. Involve others like social worker and psychologist.

Protocol for Dealing with the Angry Patient

Dealing with Angry/Upset Person

- Rapport:

  o Greet and confirm name of the person
  o Introduce self
  o Calm the person and take him/her toquiet room
  o Reassure that the problem will be discussed and explained
  o Invite a colleague or a social workerto attend.

- Set the scene:
- A quiet room, invite person to sit, and offer water, drinks
- Apologise that she/he is upset
- Ask 'can you please tell what the problem is?'
- Listen and understand the problem. This can be either person not satisfied with the care (ward, ICU, ED), delay in offering care (ED or OPD), or a mistake has occurred (wrong treatment or procedure).
- Actions:

  A. Apologise again and reassure that the problem will be solved.
  B. Use good language
  C. Do not clash or defy the person
  D. Check all the facts from file or staff
  E. Explain what has been done
  F. Show person the necessary documents
  G. Give person the chance to express their opinion
  H. Set a clear plan to deal with the problem and to discuss it with senior staff.
  I. Reassure person that everything possible will be done to sole the problem.
  J. If not satisfied, call a senior staff or administrator

- Closing:

Agree on a clear plan and wish him/her all good health.

References

Elder, N, Ricer R., Tobias B. How Respected Family Physicians Manage Difficult Patient Encounters. *Journal of the American Board of Family Medicine*. 2006, 19:553–41.

Krebs, E. E., J. M. Garrett, T. R. Konrad. The Difficult Doctor? Characteristics of Physicians Who Report Frustration with Patients: An Analysis of Survey Data. *BMC Health Services Research*. 2006, 6:128.

Hass L. J., Leiser J. P., Magill M. K., Sanyer O.N. Management of the Difficult Patient. *American Family Physician,* 2005 15; 72(10): 2063–2068.

Hull S. K. and Broquet K. How to Manage Difficult Patient Encounters. *Family Practice Management,* June 2007

Wasan, A. D., Wootton J., Jamison R. N. Dealing with difficult patients in your pain practice. *Regional Anesthesia and Pain Medicine,* 2005; 30: 184–192.

Silverman J., Kurtz S.M., Draper J.(1998)Skills for Communicating with Patients. Radcliffe Medical PressOxford

Brod T.M., Cohen M.M., Weinstock E.(1986)*Cancer Disclosure: Communicating the Diagnosis to Patients—AVideotape.* Medcom, Inc. Garden Grove CA.

Buckman R. (1994) *How to Break Bad News: AGuide for Health Care Professionals.* Papermac, London

Medical Ethics

Basic Principles of Medical Ethics

These are summarised as the following:

- **Beneficence:** Do all that is good
- **Non-maleficence**: Do no harm
- **Autonomy:** Patient's right to share decisions and plans
- **Confidentiality:** All patient information not be shared with others unless in special situations (court, public interest, approved research)
- **Honesty:** Tell the truth, never lie to patients
- **Justice:** Fairness to all

The Principle of Beneficence

In simple words, beneficence means do all good to your patients, as well as to take all necessarysteps to prevent and to remove harm from the patient. The basisof doing goodcan be applied both to individual patients and to the society as a whole. For example, the good health of a particular patient is an appropriate goal of medicine, and the prevention of disease through research and the employment of vaccines is the same goal expanded to the population at large.

The Principle of Non-maleficence

This means that one should not intentionally inflict harm to patients and persons. If one causes such an act by being careless, this can be classified as negligence. Both the law and the moral beliefs of any society, as well as ethical codes, support any notion to avoid the risk of harm to patients and community. Thus, health care professional have a duty to avoid committing medical mistakes, although such errors or mistakes can occur now and then. However, every effort should be done by health care professional to protect their patients under any circumstances.

Respect of Autonomy

Patient autonomy means that the patient has the right to share decisions, and acceept or refuse any steps of care. Respect of autonomy is thebasis for the practise of 'informed consent'. General ethics stress thatpatients and individuals should be well informed about any management plans, be it medical, surgical procedure, or tests, before asking them to sign for it. This will gurantee the basic rights of patients, and will remove the so-calleddoctor-centred medicine.

One shouldalwaysrespect the autonomy of the patient. This should be the standard practise of health care teams. One important example is that Jehovah's Witnesses who hold a belief that they should not receiveblood transfusion. Thus, if a life-threatening condition occurs to such people, the benefit of blood transfusion should be clearly explained tothem, as well as the risk of not having it. Therefore, in a life-threatening situation where a blood transfusion is required to save the life of the patient, the patient must be informed about the possibility of dying if he/she continues to refuse such treatment.

After such full explanation, the particular patient is then free to choosewhether to accept the blood transfusion in keeping with a strong desire to live, or whether to refuse the blood transfusion in keeping withhis or her religious convictions about the wrongness of blood transfusions.

What Should Be Done?

Referring to the above situation, the physician had a *prima facie* duty to respect the autonomous choice of the patient, as well as a *prima facie* duty to avoid harm, and to provide a medical benefit. In this case, informed by community practise and the provisions of the law for the free exercise of one's religion, the physician gave greater priority to the respect for patient autonomy than to other duties. However, some ethicists claim that in respecting the patient's choice not to receive blood, the principle of non-maleficence also applies, and must be interpreted in light of the patient's belief system about the nature of harms—in this case a spiritual harm.

Informed Consent

Follows from the principle of patient autonomy, and consent is required before you may provide care.

'No medical intervention done for any purpose, whether diagnostic, investigational, cosmetic, palliative, or therapeutic, should take place unless the patient has consented to it.'Informed consent also serves as a significant protection to you against possible litigation.

Consent may be expressed or implied; the former (e.g., via a signed consent form) typically occurs in hospitals and relates to specific procedures. Consent may be given verbally, but a consent form provides evidence of consent. Is not a contract, however, and the patient can withdraw consent at any time.

For routine procedures, such as a blood pressure check, consent may be implied if the patient comes voluntarily to the doctor's office for a check-up. For consent to be 'informed', the patient must receive a full description of the procedure, its risks and benefits, the prognosis with and without treatment, and alternative treatments. The patient must have the mental competenceto comprehend the information, and must give specific authorization for the doctor to proceed with the plan. The burden is not exclusively on the doctor. The patient should ask questions when they are uncertain, and should thinkcarefully about their choices.

When the patient or substitute decisionmaker is unable to consent and there is demonstrable severesuffering or an imminent threat to the life or health of the patient, a doctor has the duty to do what isimmediately necessary without consent, but by better involving two other senior colleagues.

Emergency treatments should be limited to those necessary to prevent prolonged suffering, or to deal with imminentthreats to life, limb, or health. Even when he/she is unable to communicate, the known wishes of thepatient must be respected.

Confidentiality

This forms a cornerstone of the doctor-patient relationship. It implies respecting the patient's privacy, encouraging them to seek care, and preventing discrimination on the basis of their medical condition. In order to protect the trust between doctor and patient, the physician should not release personal medical information without the patient's consent. Like other ethical duties, however, confidentiality is not absolute. It can be necessary to override privacy in the interests of public health, as in contact tracing for partners of a patient with a sexually transmitted disease. Note that you are legally obligated to report a possibly HIV infected patient to the public health authorities. However, this should always be done in a way that minimizes harm to the patient.

Scenario: A patient's relative gives you information on the patient, but asks you not to reveal where the information came from. Do you have to keep this secret?

Discussion: For the patient to be well informed and to make informed choices (i.e., autonomy), the doctor must disclose information that is materially relevant to the patient's understanding of their condition, their treatment options, and likely outcomes. This would include, for example, information on medical errors made in their care. As the American College of Physicians says, 'Errors do not necessarily constitute improper, negligent, or unethical behaviour, but failure to disclose them may.'

Scenario: A teenage patientwas brought by his friends in a confusional state after using illicit drugs.

Afterrecovering in the emergency department, he asks you not to tell his parents. How do you balance protection ofthe patient's confidentiality against the rights of her parents?

Discussion: There are some circumstances under which you may choose not to disclose information to a patient includingwhen the patient specifically asks not to be told (you should still offer them the chance to know the truth), when a patient is incapacitated (here you typically inform the family), during an emergency when the patient's condition is unstable and immediate care is required, and the controversial notion of 'therapeutic privilege', which means that the doctor deems that the risk of informing the patient is worse than not doing so(e.g., they might attempt suicide or refuse necessary treatment).

A Case-Based Approach to Ethical Decision-Making

Adapted from A. R. Jonsen, M. Siegler, W. Winslade, <u>Clinical Ethics</u>, 7th Edition. McGraw-Hill, 2010

Medical Indications

The principles of beneficence and non-maleficence:

- What is the patient's medical problem? Is the problem acute? Chronic? Critical? Reversible? Emergent? Terminal?
- What are the goals of treatment?
- In what circumstances are medical treatments not indicated?
- What are the probabilities of success of various treatment options?
- In sum, how can this patient be benefited by medical and nursing care, and how can harm be avoided?

Applying the Principles of Beneficence and Non-Maleficence

1. First of all, make sure that the patient has been well informed about the details of the problem or intervention. Also, check that the patient has understood the information given and has consented using the normal channels.
2. Second, decide, better with the aid of a committee, whether the patient is in his full mental capability and satisfies the legal competency. Check if there is any evidence of incapacity.
3. If you found that the patient is in his full mental capacity, check the stand of the patient regarding his choices and preferences of treatment.
4. If it has been decided by a committee that the patient is mentally incapacitated, then seek information from relatives or social worker whether the patient has expressed his preferences before when he/she was in a good mental state.
5. It is important to find the right person to act and take decisions for this mentally incapable patient. This is a legal step.
6. If you sensed that the patient cannot cooperate with the medical team, then find the cause and involve a psychiatrist or a psychologist.

Maintaing the Quality of Life

This is an extremely important issue. We have to strive to maintain the quality of life of patients with or without treatment and according to the principles of ethics such as beneficence, non-maleficence, and respect of autonomy.

There are some difficul questions to ask here:

*   Are there any chances that the patients will enjoy normal life with or without treatment?
*   Is it likely thatthe patient will experience any physical, social, or mental changes even if he is successfully treated?

- Can the evaluator eliminate anybias when dealing with such situations?
- Are there any ethical considerations when dealing with the qualityof life of patients who are treated or denied treatment?
- In case of forgoing life-sustaining treatment, what will be the rationale?

Applying the Principles of Justice and Fairness

- Consider conflict of interest when treating patients.
- Ask if there are any family members or others who will be interested in clinical decisions.
- Is patient confidentiality will affect legitimate interest of others?
- Consider financial aspects that may affect conflicts of interest when making a clinical decision.
- Think of any legal or religious issues that might affect the clinical decisions.
- Consider medical research issues that might affect decisions.
- Canpublic health safety be endangered by clinical decisions?

Ethical Considerations

Practical Exercise

Problem 1:

Patients who depend on special machines to keep them alive: if it became clear that such machines are not improving the condition of the patient or upon family request to switch them off, what would you do?

Example 1:

Elderly patient with COPD complicated by respiratory failure who cannot be weaned off the ventilator. The patient who can still write messages indicated his desire to switch the ventilator off.

What will be your action?

- First, you have to judge the mental capacity of the patient by a team.
- If the patient can communicate by writing, find out the reasons and the concequences for his request, and whether he has discussed this with his wife or a close relative.
- You can request a psychiatrist to see him if you are concerned about severe depression or mental incapacity.
- Discuss the issue with the wife, if available, or other close family members.
- If the patient is judged to be in a reasonable mental state, and the family agreed, then the treating team can respect the wishes of the patient and switch off the machines.

When is it justifiable to discontinue life-sustaining treatments?

- If the patient has the ability to make decisions, fully understands the consequences of their decision, and states they no longer want a treatment, it is justifiable to withdraw the treatment.
- Treatment withdrawal is also justifiable if the treatment no longer offers benefit to the patient.

Do different standards apply to withholding and withdrawing care?

Some clinicians feel that it is easier to not start (withhold) a treatment, such as mechanical ventilation, than to stop (withdraw) it. While there is a natural tendency to believe this, there is no ethical distinction between withholding and withdrawing treatment. In numerous legal cases, courts have found that it is equally justifiable to withdraw as to withhold life-sustaining treatments.

What is a 'Do Not Resuscitate (DNR) or Do Not Attempt Resuscitation (DNAR) Order?

A Do Not Attempt Resuscitation (DNAR) order, also known as a Do Not Resuscitate (DNR) order, is written by a licensed physician in consultation with a patient or surrogate decisionmaker that indicates whether or not the patient will receive Cardiopulmonary Resuscitation (CPR) in the setting of cardiac and/or respiratory arrest.

CPR is a series of specific medical procedures that attempt to maintain perfusion to vital organs, while efforts are made to reverse the underlying cause for the cardiopulmonary arrest. Although a DNAR order may be a component of an advance directive or indicated through advance care planning, it is valid without an advance directive.

The Role of Patient Autonomy

Since the original inception of DNAR orders, respecting the rights of adult patients and their surrogates to make medical decisions, otherwise known as respect for autonomy or respect for persons, has been emphasised. This concept is reinforced legally in some countries, especially in the west, which requires hospitals to respect the adult patient's right to make an advanced care directive and clarify wishes for endoflife care.

In some Middle-Eastern countries, the decision should be taken by the attending physician and two other consultants, and then discussed with the family. In general, an emphasis on improving communication with patients and families is preferred over physicians making unilateral decisions based on appeals to medical futility regarding the resuscitation status of their patients.

When can CPR be withheld?

Many hospitals have policies that describe circumstances under which CPR can be withheld based on the practical reality that CPR does not always provide direct medical benefit. Two general situations justify withholding CPR:

1. When CPR will likely be ineffective and has minimal potential to provide direct medical benefit to the patient.
2. When the patient with intact decision-making capacity or a surrogate decision maker explicitly requests to forgo CPR.

What if the family disagrees with the DNAR order?

Ethicists and physicians are divided over how to proceed if the family disagrees with the recommendation to forgo attempting CPR.

If there is disagreement, every reasonable effort should be made to clarify questions and communicate the risks and potential benefits of CPR with the patient or family. In many cases, this conversation will lead to resolution of the conflict. However, in difficult cases, the issue should be raised to the higher ethics committee.

DNR Ordersin Emergency Situations and Surgery

It is common to have patients present for surgery for which a 'DoNotResuscitate' order is written in their chart. Physicians and patients alike suffer from misconceptions about the potential benefits and harms of resuscitation in the Operating Room (OR), and even the definition itself of resuscitation in the OR requires clarification prior to surgery. Because the OR environment presents patients with a situation in which CPR carries significantly different risks and benefits than on the medical ward, rediscussion of the implications of the DNR order are necessary.

Even in emergencies, physicians have an ethical obligation to recognise and respect patient autonomy. Whenever possible, physicians should obtain input from the patient, or when the patient is incapacitated, from appropriate surrogates regarding the status of the patient's DNR orders in the OR. In the absence of such input, consensus should be reached among the caregivers about the medical benefits or futility of CPR. In any case, medical care of the patient in the absence of patient input should be directed towards realising, to the best of the physician's ability and knowledge, the patient's goals.

Practical Exercise:

Mr B. is a 68-year-old man with a history of severe coronary artery disease, peripheral vascular disease, and stroke. He suffers from right hemiplegia and mild expressive aphasia. He is awake and alert and presents for right below the knee amputation (BKA) for vascular insufficiency. His chart carries a DNR order. In the holding area prior to surgery, the anaesthesiologist discusses the DNR order with Mr S. who appears depressed. Mr S. states unequivocally that he does not wish CPR in the OR regardless of its cause or positive prognosis. He tells his anaesthesiologist that he is willing to go 'so far and no more'. The patient agrees to subarachnoid anesthesia (spinal block) and sedation. He is not intubated. After about twenty minutes, the patient complains of weakness in his arms and difficulty breathing. Within three minutes, his blood pressure and heart rate fell, and he abruptly arrests.

The decision: Should the patient be intubated? Should CPR be commenced?

It is hard to argue ethically for the institution of CPR in this patient who, while neurologically impaired, appeared to have full capacity to understand and make decisions regarding his own medical care. Despite preoperative discussion, which included information about the good prognosis from CPR in the OR, the patient stated clearly his wishes to not be resuscitated if an arrest occurs. Instituting CPR in this patient because the cause of arrest is anaesthetic-related would be like justifying transfusion in a Jehovah's Witness against their will because the surgery was the cause of life-threatening hemorrhage, yet adhering to their wishes if hemorrhage was due to non-surgical injuries.

Medical Errors

Errors are inevitable in the practise of medicine. Sometimes, these result from medicine's inherent uncertainty. Occasionally, they are the result of mistakes or oversights on the part of the individual provider. In either case, a physician will face situations where she must address mistakes with her patient.

Admitting the Error

Physicians have an obligation to be truthful with their patients. That duty includes situations in which a patient suffers serious consequences because of a physician's mistake or erroneous judgement. The ethical nature of the relationship between a physician and patient requires that a physician deal honestly with his patient and act in her best interest.

If a physician believes there is justification for withholding information about medical error from a patient, his judgement should be reviewed by another physician and possibly by an institutional ethics committee. The physician should be prepared to publicly defend a decision to withhold information about a mistake from the patient.

It has been shown that patients are less likely to consider litigation when a physician has been honest with them about mistakes. Many lawsuits are initiated because a patient does not feel they have been told the truth.

Physician-Assisted Suicide and Euthanasia

The topics of easing the dying process, avoiding needless pain and suffering, and avoiding unwanted treatments often lead to discussion over the distinction between assistance in the dying process and assisting suicide. Physician-assisted suicide occurs when a physician provides a medical means for death, usually a prescription for a lethal amount of medication that the patient takes on his or her own. In euthanasia, the physician directly and intentionally administers a substance to cause death.'(American College of Physicians, 2005)

Different constituencies have made differing laws concerning physician-assisted suicide. For example, the American College does not support its legalization, while Oregon does allow physician-assisted suicide. In Canada, it is illegal to assist a person in committing suicide (even though suicide itself is not illegal).

Religious aspects: Some religions, especially Islam, clearly state that killing or assisting in killing a person without a definite Islamic court ruling is a big sin and should not be allowed.

Summary of How to Approach an Ethical Problem

Several groups have proposed frameworks that help you to address the ethical aspects of a difficult situation in a systematic manner.

The following steps in reaching an ethical decision have been proposed:

1. Describe the case simply, but with the pertinent facts
2. Specify the ethical dilemma
3. What alternatives do you have?
4. List the key considerations: autonomy, beneficence (what are the medical alternatives?), justice (rights of patient vs family, etc.), context (situational factors such as your own feelings, your peers, the law)
5. Propose a resolution
6. Review this choice critically, formulate it as a general maxim and review its plausibility
7. Do the right thing!

Practical Communication and Ethical Issues

The use of experimental unapproved drugs

Should people who are terminally ill have the right to use whatever drug they may want in order to have a shot at saving their lives?

This comes up again and again for people dying of cancer. It also can come up for people in late stages of diseases like Parkinson's and Alzheimer's, and even for people who are becoming blind from macular degeneration. There are many, many circumstances and conditions where people run out of what contemporary medicine can give them, and want a chance at something that may be showing promise in animal studies or in early human trials. Should they have the right to get these treatments regardless of what the US Food and Drug Administration (FDA) or anyone else thinks?

1. **The ventilator off:**

   J. H., a 15-y-old girl, went in for a tonsillectomy to help with her sleep apnea. She was obese and having problems with sleep, and as often happens, the physicians there recommended that she have her tonsils out, have some tissue removed to open up the airway. Something went terribly wrong. She bled a lot. She had a heart attack. Her brain hemorrhaged.

   The doctors examined her and determined, according to braindeath protocols, that she was brain dead. They came out and told the family that she was brain dead, and the decision now had to be made about removing life support.

   The family was angry, shocked, and despondent, and this was a strongly religious family. They heard 'brain death, kind of dead, sort of dead, but maybe not really dead', and then they heard the term 'life support' and said that they didn't want doctors to take life support away from their little girl.

2. Mr G. is a 67-year-old mechanic who works as a mechanic and supports a large family. He presented to his doctor with a chief complaint of weakness. On examination, he has pallor, and he had occult blood in his stool. Laboratory tests revealed irondeficiency anemia. He underwent colonoscopy, which demonstrated a mass in the descendingcolon. The biopsy confirmed the diagnosis of Adenocarcinoma.

3. H. A., a single businessman with frequent travels to China, has recently observed that he lost 8 kg during the last four months. Initial CBC and biochemistry were within normal. When referred to the medical outpatient, atest for HIV was requested, and the result came reactive. You are required to discuss this result with him:

   • The simulator should be well instructed. If asked, he should mention that he has illegal sexual practise during his travel to China and India.
   • Also, the simulator should:
   • Question the result (disbelief)
   • Show emotional reactions and feeling of guilt
   • Should ask about the chances of cure
   • Ask about future marriage plans
   • Candidate should mention that he will follow the SPIKES protocol
   • The interaction can be in Arabic or English, but the discussion by examiners is conducted in English
   • Evaluation by using Elbagir's universal score

4. **Explaining treatment and counseling**

   O. M. is a 38-y-old lady who was discovered to have end-stage renal failure requiring haemodialysis and then possible transplant. However, the patient initially did not accept having haemodialysis thinking that it is bad for her.

Please discuss this issue with the patient to persuade her to accept treatment.

Instructions to simulator should include:

- Mention that haemodialysis will not treat him and is bad for his body
- He has seen others patients who died following this procedure
- Show emotional reactions of depression and despair
- Finally agree for a six-month period

The candidate'sperformance:

- Should explain the procedure in simple words
- Answer the patient's questions in a professional manner
- Deal with emotions and queries raised by the patient
- Stress the seriousness of not refusing this treatment

5. This obese 58-year-old woman has DM for fifteen years. She is not well-controlled because her diet contains high carbohydrates. Shealso spends most of the time in front of the TV. Her FBS is 165 and HbA1C is 9. You are asked to discuss this problem with her to help improve her diabetes control.

6. Ethics: discussion with daughter of a woman who was about to be discharged, following urosepsis. She had a previous stroke and the family wes not coping. Wanted candidate to tell the mother she had to go to a nursing home. Also, on digging further, there were marital problems, no holidays for years, etc. I offered that we should discuss with the mother together and be honest. Offered options are: OT/package of care, respite support if required.

7. **Poor glycemic control:**

58-year-oldman with T2DM had recent admission with claudication pain and angiogram revealed diffuse disease not for intervention. Had uncontrolled HT, impaired renal function, peripheral neuropathy, very high HbA1C, high cholesterol, smoker 20/d, alcohol 30u/week, not exercise regularly, care for disabled wife, poor compliance with medicines and diet, occasional hypoglycaemia.

Explore the patient attitude counsel and plan action.

CHAPTER 2

The Art of History Taking

Remember these pearls:

One should not increase, beyond what is necessary, the number of entities required to explain anything.

One should always choose the simplest explanation of a phenomenon, the one that requires the fewest leaps of logic.

Don't make unnecessarily complicated assumptions.

Make things as simple as possible, but no simpler. - Albert Einstein

Taking the history of a patient is the most important tool you will use in diagnosing a medical problem. Medical historytaking is one of the first clinical skills medical students are taught. All medical school graduates are expected to be competent in effectivelytaking a full medical history. Almost about 70–80 per cent of conditions can be diagnosed by a good medical history alone. This includes peptic ulcer, early diabetes, and angina.

The Trouble with Physicians

In one study, physicians did not allow patients to complete their opening statements 69 per cent of the time. The mean time until the first interruption was eighteen seconds. Once interrupted, fewer than 2 per cent of patients went on to complete their statements.

'Data are thus very much physician-determined, skewed towards problems that are biomedical in nature. It has been proposed that current interviewing practises are at odds with scientific requirements. They produce biased, incomplete data about the patient.'

Physician-Centred vs. Patient-Centred Interviewing

Physician-Centred	Patient-Centred
Physician's Agenda	Patient's Agenda
Biomedical Focus	Symptom Focus
Physician Gathers Data	Patient Tells Story

Opening the Interview/Setting the Agenda

An Outline for Opening the Interview

Remember:

'If you listen carefully to the patient, they will tell you the diagnosis.' (*Sir William Osler*)

How:

The need to phrase questions appropriately to the patient's age, education, culture, language, and mental state. Some patients will have no idea what you mean when you ask them 'when did youlast open your bowels?' Make sure you can rephrase such questions without causingoffence. Only a minority of patients are 'poor historians'. These mainly includethose with altered cognitive processes (e.g., through dementia) or those with lowIQ. In such cases, learn to elicit a medical history from friends, relatives, or otherthirdparty members.

Questions

There are five major types of questions used in historytaking:

- *Closed question*: limited use. A question that only gives a limited choice of answer such as 'yes' or 'no'. For example, *'Do you have pain?'*
- *Open question*: recommended. A question that can be answered freely with as much or as littleinformation as the responder wants to give. For example, *'Please tell me what is troubling you today.'*
- *Probing or explanatoryquestions*: good. These are more direct questions than open questions, as they are based on information already obtained, but allow a free response. For example, *'In what way does your leg pain affect your daily living?'*
- *Leading Questions:* not recommended. Example: *'The pain you mentioned is crushing, yes?'*
- *Multiple questions:* not recommended. *'Do you have nausea, vomiting, heartburn, or difficulty in swallowing?'*

Setting the Stage

Goal: To establish a favourable context for the interview

- Greet and welcome the patient
- Know and use the patient's name
- Introduce and identify yourself
- Explain the purpose of the interview
- Ensure comfort and privacy
- Obtain permission

Common Mistakes During History Taking

Using Inappropriate Questions

You tend to get more informative history and good details if you use the right question. Most of the time, use open-ended questions to elicit more information from the patient.

For the chief complaints, ask the patient, 'Tell me about the main or most severe complaints that brought you to the hospital.' If the patient has pain, ask, 'Can you describe the nature of the pain, I mean how do you feel it and where?' Do not ask closed questions such as 'Is the pain severe?'nor 'Does your pain feels like colic?' These are closed questions and are answered with yes or no. This will not allow the patient to describe the pain actually as he/she feels it. Avoid asking leading questions such as 'I guess your pain is burning in nature, is it not?'

Also, avoid asking multiple questions that confuses the patient such as 'How is your appetiteand have lost weight or felt unwell?'

Confusing the Patient by Asking Too Many Questions

This is a bad style and will make the patient confused and unable to concentrate. To facilitate the patient to describe the complaints, ask one question at a time and pause for the answer. This will allow the patient to give you the best information. It is true that you need to finish your task in a reasonable time, but finish it well by properly managing your time.

Failure to Listen

Many patients report that some doctors do not listen well to them. Listening is the most important communication skill. If you listen well, you get more information from patients, and they will respect you and feel more comfortable to collaborate. Also, do not interrupt patients while they are talking. This will irritate them and disturb the flow of their information.

Try to calify information if not well phrased by the patient. For example, if the patient said, 'I feel awful about the whole thing.' Then you need to clarify this by asking, 'Could you tell me what you really mean by this?'

One more important hint, observe and sensepatient's feelings, fears, and concerns. Then deal with it in the most appropriate way. This will strengthen the ties with patients, and will improve therapeutic relations. Never appear hostile or indifferent. This will discourage patients to give you more details and will feel bad about it.

Talking and Talking and Talking

Too much talking is a bad habit in any situation. It is even worse during interviewing patients or relatives. As mentioned above, listen more than you talk. You will never get informative history if you waste time talking more than the patient. Avoid lecturing the patients with jargon that is difficult for patients to understand. This will give the impression that you are showing off, which is not recommended.

Providing General Information

In some exams, there are stations about counseling or explaining treatment. Here, you need to be more specific and give full information with examples. Do not use generic information such as 'You need to eat a healthy diet'. Explain what a healty dietis. Also, when explaining a specific risk for a specific disease, such as reflux oesophagitis, explain to the patient these risks such as eating big meals, eating and sleeping after a short time, and so forth.

Adding More Mistakes

Pay attention to these mistakes:

- Forget to offer a glass of water when the patient coughs, or forget to give paper tissues if the patient starts to cry.
- Failure toshow empathy when the patient or simulates has emotional break down.
- Forgot to counsel a patient about smoking, orpot smoking, another patient I forgot to counsel about smoking also.
- Forgot to offer the 'suicide contract'(i.e., make the patient promise me that he'll call his family or the office in any case of suicidal thoughts or plans).
- Not doing the CAGE questionnaire in subjects who have alcohol drinking problems. This easy to use patient questionnaire is a screening test for problem drinking and potential alcohol problems. Four questions:

Have you ever felt you should **C**ut down on your drinking?

| No |
| Yes |

Have people **A**nnoyed you by criticizing your drinking?

| No |
| Yes |

Have you ever felt bad or **G**uilty about your drinking?

| No |
| Yes |

Have you ever had a drink first thing in the morning to steady your nerves or to get rid of a hangover (**E**ye opener)?

| No |
| Yes |

Total= | 0 | **/4**

A total score of two or greater is considered clinically significant (sensitivity of 93 per cent and a specificity of 76 per cent for the identification of problem drinking), compared with GGT liver function test, which detected only a third of patients having more than sixteen 'drinks' per day. The Alcohol Use Disorders Identification Test (AUDIT) is a longer screening tool recommended by the WHO.

References:

- Ewing JA; Detecting Alcoholism. The CAGE Questionnaire. JAMA. 12 October 1984;252(14):1905–7.
- Steinweg DL, Worth H; Alcoholism: the Keys to the CAGE. Am J Med. 1993 May;94(5):520–3.
- Bernadt MW, Mumford J, Taylor C, et al.; Comparison of Questionnaire and Laboratory Tests in the Detection of Excessive Drinking and Alcoholism. Lancet. 6 February 1982;1(8267):325–8

Summary of Mistakes of History Taking Process

- Poor Introduction
- Poor communication includingeye contact and body language
- Disorganised and non-systematic data gathering
- Non-informative family and social history
- Premature closure
- Premature counseling
- Lecture counseling
- Ignoring patient concerns
- Interrupting patient
- Poor ending: no summary and poor management plan

Examples of History and Communication Stations

Every medical student begins learning clinical skills by learning how to take a history. By the time a doctor takes the postgraduate examinations, such as MRCP PACES examination, theseskills should be second nature to him or her. Then why are so many failing in this station?

The first thing to remember is that most of the stations involve simulated patients or relatives. They have been told what to say regarding a certain condition or situation. If one asks a question, they will answer according to the instructions given. If they are not sure, they will give you the answer rather than risk hiding an important fact and prejudicing the candidate's chances.

What are the Questions One Should Ask?

The questions to be asked in history taking are standard, and these should not present a problem (presenting complaint, past illnesses, drug history, family, and social hstory, etc).

The questions that one should ask to make the history more relevant and display maturity on the part of the clinician are <u>the extra questions that are not yet considered standard.</u>

These questions are regarding the beliefs, expectations, anxieties, or concerns of the patient. If these are not asked, very important information is missed out, and this is usually the cause of failure, especially in the communications and ethics station.

If one does not take into account the thoughts and views of the patient or the concerned party, then the explanation will lack focus on the situation in, and will simply be a general explanation, which may not suffice in that particular situation.

The second part of these stations is delivering an explanation to the patient or concerned party regarding the situation. This explanation is best given by telling them in simple language what one's own beliefs, expectations, and concerns are regarding the situation. By having such a framework for assessing and explaining, the whole process becomes methodical, and thus simplified.

Consider the following scenarios:

A scenario concerning a pregnant woman who has a deep vein thrombosis—the concern of the patient is that the treatment will cause harm to the foetus. If this concern is not elicited and addressed, the explanation would be deemed unsatisfactory.

Similarly, a patient with a stroke and the scenario is regarding feeding, the relation may be concerned that not feeding and starving the patient may cause distress, or on the other hand, the concern may be that feeding would prolong the patient's suffering. Hence, it is important to elicit these thoughts and views and properly address them.

A Model for History Taking from Adult Patients

History taking is the most important part of clinical evaluation. It is a core competency in medical education and a vital part of the physician-patient encounter. It helps lead to the final diagnosis in about 70 per cent of the time. However, interviewing is also one of the most difficult clinical skills to master, and students need clear guidance to overcome this problem.

There are some minor differences in taking history as a result of teaching variations among staff related to their medical school protocols. Thus, to avoid confusion and inconsistency, it would be better to give our studentsand doctors an agreed model to adopt when taking history.

As a medical student, you often have very defined goals:

- To practise how to properly communicate and take a history
- To get information so you can identify a sensible differential diagnosis, and thus give the best care for patients.
- To give a clear case presentation to your colleagues
- To give a good impression to your consultant and colleagues

History taking is an art, which should be perfected by practising the proper methods of interviewing by adhering to the following principles:

- Strive to make the environment as private and free of distractions as possible. This may be difficult depending on where the interview is taking place. The emergency room or non-private patient rooms are notoriously difficult spots. Do the best that you can, and feel free to be creative. If the room is crowded, it's okay to try and find alternate sites for the interview. It's also acceptable to politely ask visitors to leave so that you can have some privacy.

- If possible, sit down next to the patient while conducting the interview. Remove any physical barriers that stand between yourself and the interviewee (e.g., put down the side rail so that your view of one another is unimpeded, though make sure to put it back up at the conclusion of the interview). These simple maneuvers help to put you and the patient on equal footing.

1. Establish rapport with patient. Be friendly and empathetic.
2. Indicate whether history is given by the patient or his relative/attendant when the patient is unable to communicate (comatose, aphasic).

- Use proper communication skills: listening, open-ended questions, using verbal and non-verbal skills, facilitation, and responding to patient's ques and emotions.
- Gather information in a systematic, organised, informative, relevant, and detailed way.
- In certain situations, such as emergency situations, take focused relevant history.
- As far as possible, use the patient's own words when describing the presenting complaints. Ask about positive symptoms and the relevant negatives.
- Avoid using medical terminology, which is jargon to the patient.
- Stick to the agreed sequence of different components of history taking, <u>details of which will be mentioned within each section.</u>

Steps of History Taking

o Personal information or patient profile: name, age, marital status, occupation, permenant residence, date, and route of admission. Source of history.

o   The chief (main) complaint(s) and duration
o   Background and details of the history of presenting illness (HPI): background, details, associated symptoms, constitutional symptoms, and course of the illness.
o   Review of symptoms in other systems
o   Past medical, surgical, and gynaecological history when relevant.
o   Medications and allergy
o   Family history
o   Social history: living conditions and also includes smoking, alcohol, and illicit drug use
o   Travel history

General Approach

- Prepare the scene: quiet area, maintain privacy, and patient comfort
- Rapport: greet, introduce yourself (name and rank), explain purpose of interview, and take permission.

Examples

- Hello/salam alaykum Mrs, my name is AB. I am a fourth year medical student here at the medical school. I will be interviewing you for about thirty minutes to learn what kind of problems you are having, and how they have affected you so that we can diagnose your illness and offer help. Will this be okay with you?
- Assessing the patient's comfort is the next step. An IV or oxygen mask, facial expressions of distress, or an emesis basin at the bedside provide non-verbal clues to the alert clinician. Bringing a cup of water, raising the head of the bed, or helping the patient to the bathroom may be greatly appreciated. They also provide a natural opportunity for a caring touch. Questions such as 'How are you feeling?' 'Are you comfortable now?' 'Do you feel well enough to talk now?' are helpful.

- If family members or other visitors are in the patient's room, the physician should introduce him or herself to all those present, and explain the purpose of the interview.
- Take information using the patient's own words and be neutral.
- Listen attentively and do not interrupt.
- Use Facilitation Techniques

To obtain accurate, unbiased information, exert only as much control over the interview as needed. The student's/physician's task is to keep the patient talking about the illness in a productive fashion. Facilitation techniques are employed to encourage and guide the patient's spontaneous report. These include the <u>use of posture, gesture, and words to indicate that the interviewer is interested in what the patient is saying</u>. These techniques reassure the patient that he or she should go on speaking and provide time for the patient to think and respond.

- Ask simple open-ended questions. Avoid leading and multiple ones.
- Agree with the patient on the information taken in each step.
- Ask the patient if there is anything else she/he would like to mention.

1. Personal Patient Information (Not Biodata)

Source of history and why, full name, call patient by his first or last name (e.g., Mr Ali, Mrs Smith, Ms N. In some cultures, like the Middle-East, one can use local Arabic titles such as 'uncle', 'aunt')

<u>Specific Questions</u>

Patient's name

- Please tell me your first and last name.
- Could you please tell me your first and last name?
- Can you spell your last name for me please?

Determine the patient's age, height, and weight

- How old are you?
- When were you born?
- How much do you weigh?
- What is your height?

Determine the patient's occupation:

> What do you do for a living?
> How long have you worked in your present job?
> What did you do before your present job?
> Is your work stressful?

Is there much physical activity associated with your work?

- How long have you been retired?
- Gender (sex): obvious
- Date and route of admission or clinic visit
- Date of birth or age
- Marital status
- Permanent residence and address(place where he/she stayed most his/her years)
- Occupation;looking for occupational hazards;if retired, mention his previous job(e.g., retired banker or teacher)
- Ethnic background when relevant (not tribe): May suggest some endemic diseases(e.g., Asian, African, Caucasian). Avoid white, black, or coloured descriptions.

2. Chief or Main Presenting Complain(s) and Duration.

- Ask: What brought you to the hospital or the clinic today? or: What seems to be the problem today?
- Usually one, but can be multiple. If the list is extensive and obviously beyond the time limit available for the interview, ask,

'Which of these problems concerns or bothers you the most?' or 'Which of your problems did you hope I could help you with today?'

- It might a worsening chronic problem (e.g., worsening dyspnoea for last two weeks)
- Record this in the patient's 'own words'
- Should be short and specific as described by the patient
- Example: Shortness of breath for two days
  Cough for five days

3. History of Present Illness (HPI):

This is the most important and challenging part of history taking. You have to dissect the chief complain(s) in an orderly and informative pattern.

- Start with a brief background of the patient. Example: This patient is known to have (or diagnosed as) bronchial asthma for the last five years. He was well till he . . .

  Avoid the term: known case as, ethically, the patient is a human being with a medical problem.

- If the patient has no previous illness, mention: This patient has noprevious diseases. Avoid saying medically free.

Questions should be simple and clear so that the patient has no difficulty understanding what is being asked. Avoid using technical terms and diagnostic labels.

The interviewer's questions should indicate what <u>type of information</u> is requested, but not what <u>answer</u> is expected.

Example: The difference between asking 'tell me about your stomach problems', and 'your stomach is hurting in you, isn't it?' is obvious, but it is easy to fall into the pattern of asking leading questions. Effective questions are usually simple. Avoid double-barreled questions such as 'Are you having any stomach pains or bladder problems?'

Ask also if this complaint has occurred before. Then ask about the details of the chief complains in a <u>chronological and systematic way</u>.

This should include:

1. What was the patient doing at the time?
2. Onset: <u>sudden or gradual</u>. If the CC is pain, ask about the site of maximum intensity, radiation, character (nature), severity using one to ten pain scale, and aggravating and relieving factors and course.

   You may use mnemonic <u>LORD SANFARO</u>:

   L- Location
   O- Onset
   R- Radiation
   D- Duration
   S- Severity
   A- Aggravating factor
   N- Nature
   F-Frequency
   A- Associated factors
   R- Relieving factors
   O- Offset

   Or simply follow the word: SOCRATES

   S: Sit
   O: Onset
   C: Character
   R: Radiation
   A: Associated factors

T: Time course

E: Exacerbation/relieving factors

S: Severity

Pearls

- Repeat this series of questions for each chief complaint. Ask one question at a time; avoid multi-part questions.
- Some questions won't work in certain situations. For example, fatigue doesn't have a location.
- Record the information as objectively as possible without interpretation. Avoid medical jargon unless the patient uses it.
- Quote the patient directly as needed.'My teeth itch,' for example.
- Pay close attention to the time course of the symptoms. Has symptom complex changed over time? This is particularly important with neurologic, chest, and abdominal diseases.

Example: The character of the pain of a myocardial infarction is often described as 'crushing' ortightening around the chest or 'as if someone standing on the chest'. <u>The patient's exact words are important</u>.

1. Try to use the patient's own words if possible. Some patients use highly descriptive or emotion-laden terms like 'It felt like someone was stabbing me with a knife'. This provides important clues about the patient's emotional state and reactivity. Other patients need the interviewer's help to find descriptive language.
2. Providing the patient with a choice of descriptions (leading questions) such as 'Was the pain sharp or dull?' are not desired, but in some situations, it may be necessary, although the clinician should realise that limiting the patient's response to these two alternatives can bias the history.

   - If there is more than one complaint, elaborate on each one as above, and try to link them to each other <u>in a time</u>

<u>sequence</u>. Chronologic description provides the framework for characterising the course of an illness. The interviewer should obtain a <u>chronologic re</u>port by asking when the problem first started and facilitate a continuing flow of information with questions such as 'And then what happened? . . . and then? . . . and after that?'

Example: The patient woke up at 6 a.m. with crushing chest painetc. One hour later, he became short of breath even at rest. etc. Then half an hour later, he collapsed and was unable to stand up and go to the bathroom.

- Ask about any associated symptoms with each complaint. Symptoms rarely occur singly. The student/doctor should listen for groups of related symptoms that provide diagnostic clues about pathologic processes and involved organs.

  Example: Ask 'When you had the joint pains, did you notice anything else?' If the patient's response is positive, he or she is asked to describe the associated symptoms through open-ended questions. Further clarification can be obtained later using more specific questions.

- Record the effect of the symptoms on the daily activity of the patient, and whether he/she has missed school or work. Also, enquire about the patient's emotional reactions to the illness and the patient's means of coping with discomfort and disability. The patient's reactions to events are often as important as the events themselves.
- Ask about constitutional symptoms such as fever, appetite, non-intentional weight loss or gain if these were not mentioned earlier.
- **Course of the illness** (better than saying hospital course):

  Record any treatment or measures done to the patient <u>before</u> coming to your hospital.

Example: He was seen in the nearby primary care centre and was given IV fluids and painkillers. During stay in your hospital, mention whether there is improvement, complications, or deterioration needing special medical or surgical care. Do not include management given in the hospital such as IV fluids, U/S, and antibiotics.

4. Review of Symptoms in Other Systems

- Do not repeat the presenting symptoms.
- Enquire about symptoms related to body systems other than the presenting one.
- Be selective(i.e., focus on the system relating to the patient's problem list, and include others only if clearly related to the differential diagnosis).

Checklist for Systems Review

For each symptom, describe:

- Onset
- Duration
- Course
- Severity
- Precipitating Factors
- Relieving factors
- Associated features
- Previous episodes

General:
Fatigue/malaise
Fever/rigors/night sweats
Weight/appetite

Skin rashes/bruising
Sleep disturbance

Cardiovascular:
Chest pain
Shortness of breath (including on exercise)
Orthopnoea
Paroxysmal nocturnal dyspnoea
Palpitations
Ankle swelling

Respiratory:
Chest pain
Shortness of breath/wheeze
Cough/sputum/haemoptysis
Exercise tolerance

Gastrointestinal:
Appetite/weight loss
Dysphagia
Nausea/vomiting/haematemesis
Indigestion/heartburn
Jaundice
Abdominal pain
Bowels: change/constipation/diarrhoea/
Description of stool/blood/mucus/flatus

Musculoskeletal:
Pain/swelling/stiffness in muscles/joints/back
Restriction of movement or function
Power
Able to wash and dress without difficulty
Able to climb up and down stairs

Genitourinary:
Frequency/dysuria/nocturia/polyuria/oliguria
Haematuria
Incontinence/urgency
Prostatic symptoms
Impotence
Menstruation (if appropriate): Menarche (age at onset), duration of bleeding, periodicity, menorrhagia (blood loss), dysmenorrhoea, dyspareunia, menopause, post–menopausal bleeding

Central Nervous System:
Headaches
Fits/faints/loss of consciousness
Dizziness
Vision: acuity, diplopia
Hearing
Weakness
Numbness/tingling
Loss of memory/personality change
Anxiety/depression

Endocrine:
Menstrual abnormalities
Hirsutism/alopecia
Abnormal secondary sexual features
Polyuria/polydipsia
Amount of sweating
Quality of hair

Skin
Rash
Pruritus
Acne

Summary: At this stage, give the patient a summary of the HPI. Say. 'Before we go on, let's see if I understand your history. You were quite well till two days ago when you first noticed . . .' This summary gives the patient a chance to check the accuracy of the history and gives the student/physician a chance to review the history for gaps or lack of clarity.

5. Past Medical and Surgical History

- Major elements of the past medical history include childhood and adult illnesses, operations, trauma, allergies and drug sensitivities (characterised in detail), and immunisations.
- Ask the patient whether he has any <u>relevant past medical illnesses, which could be related to the current presentation.</u> Example: If the patient presented with epigastric pain and vomiting, ask about previous history of gastrointestinal diseases such as peptic ulcer or gall bladder problems.

  If a patient presented with shortness of breath, ask about past history of cardiac or respiratory diseases such as ischemic heart disease, COPD, or bronchial asthma.

- Explore any <u>risk factors</u> for the current problem. Example: If the presentation is jaundice, enquire about risks for hepatitis such as blood transfusion, sharing needles, or dental visits.
- Enquire about presence of major illnesses such as DM and hypertension.
- Ask about any previous surgeries, as they might be related to the presentation.
- In women, ask about any relevant gynaecological history.

Surgical History

Ask about any previous major operations or surgical procedures and the indications for these surgeries. Also, ask about any complications that occurred.

6. Medication history

- Please tell me if you are taking any medications.
- *If you know the dose, tell me.*
- How often do you take your medications?
- Do you take oral pills or injections?
- Tell me if you are allergic to any substance or drugs.

Ask if anything happens to them when they take the drug. Sometimes, the patient may be intolerant to the medication. However, be aware of rashes, swelling, and other signs of anaphylaxis.

- Provide a list of all drugs taken <u>before</u> presentation. If the patient is unable to remember the names, ask her/him to provide the drug boxes or leaflets.
- Record medications using the generic names (trade names in brackets), doses, and duration of treatment.
- Also, ask about use of herbal and traditional medicines.

7. In Women, Ask About Obstetric and Gynaecological History

Past obstetric history

- If married: Do you haveany children and how many?
- Were the pregnanacies full term, and how were you delivered?
- Did you have any complications?
- Were your children all healthy, any abnormal babies?

Past gynaecological history

- Have been told to have any problems with your reproductive organs?
- Any surgery performed on you?
- If relevant, any recent cervical smear

Family History

- Medical problems in family members should be reviewed with special attention to inherited disorders. Ask first about the father, mother, brothers, and sisters whether they are alive and well. If there is death in the family, ask, if possible, about possible causes.
- Then ask if there is a similar problem (like the presenting one) in the family.

Example: In a young patient presenting with jaundice, ask about family history of such problem, putting in mind theinherited conditions such as Wilson's disease, haemochromatosis.

- If a familial disease is found, try to make a family pedigree so as to have an idea about the mode of inheritance.

Sample Family History

Use the following notations:

Male	Female	Deceased	Affected Age
□	○	◿	⊛46

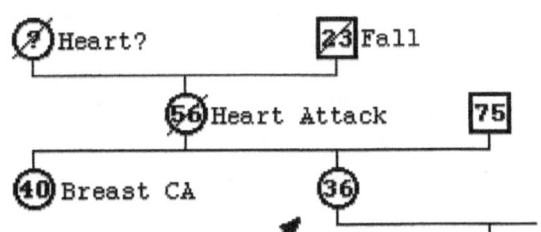

Fig 1: Skech sample of family history

The sample shows a 36-year-old female, married with one female child. The patient has a living sister with breast cancer, and her mother died of a heart attack at age fifty-six.

9. Social History

- This section is very important and often ignored by students and doctors.

  During the social history portion of the interview, the student/physician can gather data about the patient's living conditions and housing.

- Determine if they live alone or with others in a house, flat, sheltered housing, residential, or nursing home. Is the house/flat owned or rented?
- Also, education, usual daily activities, functional status, relationships with friends and family, social supports and stresses, financial status/insurance coverage.

Functional Status Questions: The effect of illness on daily activities

This is very important. Use the following questions:

Tell me how are you coping athome?You can ask more details as below.
Tell me if you have any difficulties in performing daily activities like bathing, dressing, or cooking.
Are you able to move around your house?
Is there any need to use a wheelchair or frame or walking stick?
If you live upstairs, is it easy to climb stairs?
Tell me if you ever needed a home help

- Thenask about habits such as use of cigarettes, alcohol, and illicit drug use that have known health consequences. Inappropriate or extramarital sexual activities can be asked for if there is suspicion of sexually transmitted diseases; in particular, HIV or hepatitis.
- Asking about such sensitive issues in a conservative society should be conducted with care using proper communication.

This includes first apologising then explaining the purpose. Also, maintain and assure the patient about privacy and confidentialityand take permission.

Example of questions:

Please forgive me. I would like to ask you further questions regarding some sensitive issues. The questions will be very personal, but please don't feel embarrassed, and you don't have to answer them if you do not wish to. The purpose of such questions is just to gather information to help us reach a diagnosis.

We usually ask these questions to all patients. If the patient accepts, then mention that thesequestions refer to cigarette smoking, alcohol consumption, and sexual habits. You have the right to answer or decline.

- **Smoking History**

  Have they ever smoked cigarettes? If so, how many cigarettes or packs per day, and for how many years?

  If the patient mentioned that he/she stopped smoking, ask when did this occur? The packs per day multiplied by the number of years gives the pack-years, a widely accepted method for smoking quantification. Pipe, cigar, and chewing tobacco use should also be noted. Also, in some communities, ask about shisha smoking. This is a water pipe device that delivers nicotine through the tubing device. It carries the same risks as cigarette smoking.

- Also ask about alcohol and be care in certain communities and cultures, so repeat apology. If the answer is yes, then ask how much alcohol in an average week. Also, for how long. It is difficult to know different types of alcohol, but if possible, ask about the type of alcohol. This will give an idea about the damage that may have been caused.

Sexual History and Practices

In conservative societies, ask if there is suspicion of HIV, sexually transmitted diseases, or any disease that is transmitted by sexual contact such as hepatitis and Reiter's syndrome. The key questions to ask to ascertain risk to the patient and their partner are:

- *First, ascertain whether the person issingle or married.*
- *More details: Do you practise sex?If yes, with a single or multiple partners?*
- *When you had this problem, have you practise casual or regular sex? Also ask if the person has male or female partners.*

10. Travel History

1. Will be relevant if there is suggestion of infectious diseases such as HIV, hepatitis, malaria, kalazar, brucellosis, enteric fever, diarrhoeial diseases, or haemorrhagic fevers.
2. Travel can be within the same coutry or outside it.
3. You need to have an idea about the geography of such diseases.
4. So ask specifically about the place you travelled to, and relate this to the possible connection with the medical problem.

Psychiatric Historyand Mental Health

This is a specialised area and relates to persons with mental illnesses. A psychiatric history is different from the mentalstate examination, which determines the actual mental state of the patient at the time.

Start off with the usual protocol of historytaking:

- Full personal information as above.
- Is history taken from the patient himself or a relative, and why?
- Enquire about the main presenting complaint.

Changes in mood

- *Do you feel very low on the past few weeks?*
- *Have you lost interest in doing what youused to enjoy previously?*
- *Please tell me about your sleep pattern (early morning waking or sleeping in)*
- *How is your appetite?*
- *Do have same interest in family life and sex?*

Deliberate self-harm

- *Have you ever had thoughts about hurting yourself or ending your life?*
- *Have you done so?*
- *Have youhad any suicidal ideas?*
- *Tell me if you feel that your life is worth living or not.*

Past medical history

- *Any previous chronic illnesses that may have affected your life?*
- *Have you been offered any treatment?*

Drug history

- *Tell me if you are taking any prescribed medication at the moment?*
- *Are you allergic to any substance or drug?*

Recreational drug history

- *Have you ever used recreational or illicit drugs?*
- *If yes, what are these?*
- *How often do you use them?*
- *Then ask about alcohol and smoking as illustrated above*

When asking about sensitive issues, follow the correct communication skills. First, apologise, then explain the purpose of the question, and then get permission from the person.

Milestones or life history

Find out about the patient starting from their early childhood all the way to the present day.

Overview of the UK Mental Health Act 1983 (As Amended by the Mental Health Act 2007)

The 1983 Mental Health Act had two aims: to give specific rights to patients who are, or appear to be, suffering from a mental disorder, whilst at the same time allowing for their compulsory detention and treatment.

Part II of the Mental Health Act deals with the compulsory admission to hospital for patients not involved in criminal proceedings. Short-term compulsory powers can only be used if the individual is suffering from a mental disorder of a degree.

A mental disorder is defined as follows: any disorder or disability of the mind. This includes, but is not limited to, affective disorders, such as depression or bipolar disorder, schizophrenia, neurotic, stress-related and somatoform disorders; organic mental disorders, such as dementia and delirium; personality and behavioural changes brought on by brain injury; personality disorders and mental and behavioural changes caused by substance and/or alcohol misuse; eating disorders; learning disorders (as long as there is attendant abnormally aggressive or seriously irresponsible conduct for s.3); autism. Alcohol or substance misuse does not, of itself, constitute a disorder or disability of the mind.

Closing the Interview

Before closing the interview, ask the patient if there is anything else he or she would like to discuss, or if there are any questions or explanations. This will provide an opportunity to find some information that was forgotten by the patient.

Also, summarise the information gathered andagree in a management plan that should include follow-up. As a medical student, you may not have the knowledge or the authority to construct a management plan; however, you can still practise this final part of the history according to what you have been taught. Tell the patient what will happen next. For example, 'I will get the specialist doctor to come to talk to you now' or 'You may need further tests, and these will be explained to you by the specialist'.

Always check that the patient has understood what you told him/her, and that the personagrees with the management plan. End up by saying, 'Hope we have discussed all issues that you need to discuss.' Then thank the patient and wish her/him all good health.

Some problems or difficulties may arise during the medical interview. These can be roughly divided into three categories:

1. Problems with the patient

    You may encounter some patients who are talkative, vague, silent, aggressive, anxious, or sevrely emotional. You have to do ypur best to deal with such difficult situations with the help of your team.

2. Shorcomings of the interviewer or doctor

    This mainly refers to:

    - Poor communication skills
    - An inappropriate orjudgemental attitude

- The use of inappropriate questions such as more closed-ended and leading questions
- Failure to applylistening skills
- A poor time managemnt: the habit of rushing to finish the task. This is a bad habit. This leaves a negative impression on the patient.

3. Barriers: these are mainly communication barriers such as language, culture, and personal attitiudes.

Try to avoid or deal appropriaely with these problems.

Example of Presenting the History

A. S., a 57-year-old right-handed married driverwho lives in town, was admitted to the hospitalbecause of the sudden onset of slurred speech and left hemiparesis for the last six hours.

A.S. was diagnosed with coronary hear disease sixteen months ago, and taking multiple drugs including aspirin. He was well until two days earlier when he felland fractured his left humerus. His daily aspirin was discontinuedbecause of the fracture. He was last seen well by his mother-in-lawat 2 p.m. At 2.15 p.m., she noted that he was drooling, hadslurred speech, and could not swallow pills. She called hiswife, who came home from work immediately and found him unableto move his left arm or leg. He stated that he did not havea headache, nausea, or neck pain. He was brought to the hospitalby ambulance at 4.30 p.m.

The patient had a history of coronary and bilateral carotid artery atherosclerosis. Combined coronaryartery bypass grafting and left internal carotid endarterectomy had been performed sixteen months before admission. He had asymptomatic, mild-to-moderate stenosisof the right internal carotid artery, well-controlled hypertension, hyperlipidemia, degenerative arthritis, and borderline diabetes, and was obese. His medications were fluvastatin, irbesartan, ibuprofen, aspirin, and acetaminophen with oxycodone as neededfor the recent fracture. He was allergic to penicillin. He livedwith his wife, and was employed as a civil servant. He had smokedand consumed alcohol in moderation in the past, but had stoppeddoing both at the time of coronaryartery bypass grafting. Therewas no history of atrial fibrillation, clotting disorders, cardiacvalvular disease, or use of illicit drugs.

Examples of History Taking

- **This 51-year-old patient (AB) who works as an accountant complained of chest pain. Please take relevant history.**

  How would you proceed?

- Greet, introduce self, confirm patient name, explain the purpose, and take permission
- Confirm the chief complaint and duration: Mr A. B., can you please tell what brought you to the hospital?
- A. B.: Since last night, I felt a compressing pain under the breastbone (refers to the sternum).
- Proceed to history of the present illness. Start with a brief medical background of the patient. Mr A.B., have you been suffering from any previous diseases?
- A. B.: Well, I have been diagnosed to have high blood pressure since fifteen years, and my doctor put me on two pills.
- Then ask about details of the chest pain and anyassociated symptoms: Mr A. B., can you please tell me what actually happened to you?
- A. B.: In fact, I have been suffering from chest discomfort since the last three months. I usually feel it whenever I get angry or I walk upstairs. Last night, we had a heavy dinner with friends, and after midnight, I woke up with severe chest pain. More severe than I ever had. My wife noted that I sweated a lot. So we came to the emergrncy department.

- Doctor: Please continue and tell me how did you feel this pain, and if anything that eased it.
- A. B.: Like I said, it was a compressing pain under my breastbone and moves to my jaw and left arm. I had to takeit easy, stop any heavy work.
- Doctor: Tell me if you experienced any other any other complains beside this chest pain.
- A. B.: I felt a bit dizzy, and my heart was beeting rapidly. Is that what you asked about?
- Doctor: Yes, but were you short of breath at the time?
- A. B.: Yes, I forgot to mention this. I had to open the window to have a fresh breathe. I also sweated a lot.
- Doctor: Tell me about your weight, was it steady?
- A. B.: Oh doc, I have been adding more weight. This I could't control.
- Doctor: Mr A. B., please tell me if you have any of the following complaints. Then ask about: nausea, vomiting, cough, headache, fever, dizziness, heart racing, jerks, or joint pains.

Past medical and surgical history: Mr. AB, you told me that you are under treatment for high blood pressure. Besides that, were suffering from DM or high lipids (answer was no).

- Have you had any previous surgeries? (Answer, No)
- Then complete history and ask about family diseases, social history.
- Doctor: Mr A. B., who are you living with?
- Iam living with my wife in our owned apartment.
- Doctor: Forgive me, Mr A. B., for the following kind of personal questions. I want to ask about smoking and alcohol. Is that okay?
- A. B.: No problem, doc, I have been smoking two packs a day for the last eighteen years. I drink one bottle of beer a day.

Now, Mr A. B. asked you: doc, what do you think is wrong with me? Here, you need to give a sound differential diagnosis and explain what will happen next. That is the management plan. Most likely, Mr A. B. had an acute coronary syndrome. Aortic dissection is propable, but less likely.

Tell Mr A. B. that you will be examined and and then an ECG, blood workup, will be performed, and the cardiologist will decide on further treatment.

- **History Taking Example 2**

You are the medical on duty, and the emergency department sister called you to see a patient urgently.

A 38-year-old female teacher is brought to the emergency department because of sudden severe headache and drowsiness. One of her colleagues reported that while she was sitting in the office, she was noticed to hold her head and then she fell on her desk. A relative of the patient arrived with her to the emergency department.

What further questions would you ask about?

- The background of the patient: before going in details, ask whether the patient was suffering from any chronic diseases before. In particular, hypertension, diabetes, and any neurological diseases.
- Then explore the details of what happened from the patient if she can give such information or from a witness. It is obvious that the onset of the problem was sudden, and we need to have answers to the following: the site of the headache, frontal, occipital, or global. Also, was it unilateral or bilateral? Sudden severe occipital headache is suggestive of subarachnoid haemorrhage while a unilateral one suggests migraine.
- Ask whether the headache was throbbing or not, any associated nausea or vomiting preceding or occurring after the headache.
- Associated neurological symptoms are very important. Ask about visual disturbances (migraine, stroke, subarachnoid haemorrhage). Any abnormal movements or jerks (seizures),

79

loss of consciousness, memory loss, incontinence, neck pain, and biting of the tongue.

- Any abnormal sensations or weakness of any part of the body.
- Cardiovascular events such as palpitations (arrhythmia), chest pain (acute MI).
- Past medical history. Ask about chronic diseases: hypertension, DM (hypoglycemia), any neurological problems or seizures, cardiac diseases AF, IHD, or trauma.
- Medications: Any cardiac(for arrhythmia, aspirin, warfarin) ormedications for neurological problems or medications for DM. Also ask about recreational drugs.

Whan the patient improved partially, she tells that the headache came suddenly felt in the forehead. It was the most severe headache she ever had and then she lost consciousness. Her colleagues confirmed the jerky movement, and she reported that she was wet (incontinance). She could not recall any chest pain or palpitations. She reports that she was otherwise fit and not using any medications. She denied any visual disturbances (photophobia, double vision).

Now the headache eased down, but she is still slightly confused.

What working diagnosis and differential diagnoses would you you suggest?

The most likely working diagnosis is subarachnoid haemorrhage.

Differential diagnoses:

- **A.** Cerebral haemorrhage (less likely, no weakness)
- **B.** Acute migraine (usually unilateral and no loss of consciousness)
- **C.** Temporal arteritis (age, unilateral)
- **D.** Cluster headache
- **E.** Cardiac problems (arrhythmia, MI)

CHAPTER 3

The Art of Physical Examination

This includes general examination and specific examination of different systems.

Common Mistakes

- Not reading the instructions carefully

  Read the task as well as the patient's problem. The verbs in the instructions matter, as do the limits stated in the questions. For example:

  o If the task is to conduct a focused history, then you will only get credit for taking a history in a clinically appropriate manner. You will not get credit for educating the patient or advising him or her when the assigned task is to take a history.

  o If the task is 'assess and advise' or 'discuss' or 'counsel', then will get credit for tasks like eliciting key information about the patient's problem, understanding how the patient perceives the problem, and for actions like advising the patient, providing information, and recommending follow-up depending on the nature of the presenting problem.

o If the task is 'assess and manage', then you will get credit for assessing the patient (e.g., relevant history and/or physical exam), and for managing the problem, which may include ordering investigations and making immediate treatment decisions.

- Poor technique and clumsiness. Also, not doing the right maneuvers for specific examinations. Examples: Not turning the patient to the left side when detecting the cardiac impulse. Also, not turning the patient to the left side when feeling for a small spleen.
- Misinterpretation of the physical signs

Examples: Not knowing what a brisk jaw jerk means or vertical nystagmus indicates. Also, poor knowledge of murmurs and heart sounds.

- Poor time management. No time to finish the physical examination.
- Missing to complete the differential diagnoses.
- Forgetting to auscultate the abdomen in abdominal examination.
- One candidate reported: When I was about to start the examination, the simulator told me could I please put my cell phone in the table. I took the phone and put it on the table, and then I did not change my gloves again!
- When I was doing abdominal examination, I forgot to pull out the footrest and noticed that at the end of the examination, I said I'm really sorry about that and the patient was having abdominal pain.

Common mistakes continued:

- Deficient general examination: not observing general look and the position of the patient in bed, or his surroundings (O_2 machines, sputum bottles, inhalers, etc.)

- Jumping to hands or main systems before reporting signs in the face, neck, and then hands. It is better to begin with face and neck before hands, as most of the primary signs such as facial look, eyes, jaundice, central cyanosis, mouth, including the tongue, thyroid, and lymph nodes are located in the face and neck.
- Poor examination techniques in eliciting physical signs in different symptoms (being clumsy and unpractised).
- Inability to correctly interpret the findings, and thus come up with a logical differential diagnosis.
- After all, the most deadly behaviour is to be shaky, unsure, hesitant, and easily changing minds and track when confronted with simple questions or challenges.
- Being either rough or non-courteous to patients (e.g., rough percussion and rough performance of certain maneuvers, and finally not observing the face for pain or discomfort).
- Remember that any rough handling of the patient that results in physical or psychological suffering such as shouting, being rude, verbal abuse, or inflicting excessive pain during examination will lead to outright failure of that section of the examination in the MRCP PACES.

The Long Case

- Most examination boards have dropped the long case because of the high subjectivity of examiners and the low discriminative value among candidates.
- However, other boards have adopted the method of observed long case examination to improve the power of this section of examination. The MRCP (UK) introduced the integrated case in station five (brief clinical encounter) where the candidate is expected to perform a focused history and relevant physical examination.
- It has also been agreed to replace the long case with a new station called "Comperhensive Patient Evaluation". This is somewhat similar to station 5 in MRCP PACES. This includes

focused history taking, physical examination, and discussing the differential diagnosis and management.

Examples of Long Cases

Usually, patients with rich and eventful history and multiple problems are selected. However, you may encounter patients who have eventful history, but lack significant abnormal physical findings.

Cardiovascular System

- Valvuler disease (e.g., mitral valve disease, aortic valve diseasecomplicated with heart failure, embolic phenomenon, or other complications)
- Ischaemic heart disease, with or without interventions, and may be complicated with heart failure, arrhythmia, or stroke
- Dilated cardiomyopathy
- Hypertensive (± diabetic) heart disease
- Constrictive pericarditis

Respiratory System

- Tuberculosis, virally chronic with evidence of fibrosis or cavitations
- Bronchiectasis
- Interstitial lung disease
- Cancer of bronchus with complications
- Pleural effusion (different causes)
- COPD, cor pulmonale

Gastrointestinal and Haematological

- Hepatosplenomegaly with or without lymphadenopathy (infections, chronic leukaemias, myelofibrosis)

- Liver cirrhosis and portal hypertension
- Inflammatory bowel disease
- Chronic abdominal pain (peptic ulcer)
- GI malignancies (e.g., hepatoma)
- Malabsorption
- Myeloproliferative diseases, lymphomas

Kidney Disease

- Nephrotic syndrome
- Chronic renal failure with different complications
- Polycystic kidney disease
- Transplanted kidney

CNS

- Hemiplegias (ischaemic or haemorrhagic)
- Paraplegias (cord compression by TB or tumour or transverse myelitis)
- Quadriplegias: high cervical lesions
- Gullain-Barre Syndrome
- Multiple sclerosis
- Peripheral neuritis
- Subacute combined degeneration of the cord
- Myopathies
- Lateral medullary syndrome

Miscellaneous

- Diabetes mellitus with complications
- Rheumatoid arthritis
- Connective tissue diseases (e.g., SLE)
- Thyroid disease (hyperthyroidism with complications, hypothyroidism)

How to Approach the Long Case

a) Complete and comprehensive history

This is the cornerstone of the long case. Why do candidates go wrong?

1) Failure to have a professional approach. Failure to introduce themselves and explain to the patients the purpose of the interview. Also to thank him/her at the end.
2) Failure to be systematic like confusing HPI with past medical history.
3) Ignoring the chronological order of events (time sequence), and not mentioning the necessary details about each symptom.
4) Not elaborating on the family and social history and environmental surroundings of patients.
5) Poor language and communication skills. Examples:

   a. The use of close-ended questions (e.g., do you have constipation?), instead of open-ended ones (e.g., tell me about your bowel habits).
   b. Not giving the patient enough chance to express himself in his own words or to voice out his fears and concerns. Also, by repeatedly interrupting the patient.
   c. Some candidates ignore the patient by failing to adopt eye and verbal contact with the patient.
   d. Poor communication when asking about smoking, alcohol, or sexual practises.

6) Inability to crisply summarise the history and then formulate a problem lists or differential diagnoses. This is usually feasible if the history is informative and complete.

In summary, to be a good doctor, remember the following:

1. Be organised and systematic.
2. Keep your confidence and adopt a professional (rather than a haphazard) approach.
3. Follow the correct communication methods during history taking and communication.
4. Perform physical examination with utmost skills, use proper techniques, and avoid clumsiness and roughness.
5. Correctly interpret your findings, and be able to give a sound differential diagnosis.
6. Be rational and adopt clinical reasoning and logic during discussion. Clearly support (and defend) each diagnosis you suggest with the available strong evidence derived from the history and examination.
7. Do not jump to conclusions and avoid snap diagnosis.
8. Be prepared to give a sound management plan that includes:

   a. The immediate useful investigations starting with blood, urine, and sputum followed by non-invasive tests whenever relevant.
   b. Suggest the most helpful imaging tests (e.g., US, MRI, CT, angiography)
   c. Avoid suggesting invasive techniques if the yield will not be adequate or helpful for managing the patient.

- Concerning patient management, start with general and supportive and emergency treatments such as resuscitation (IV fluids, blood transfusion, and correction of electrolytes). Then suggest specific drug treatment, if any, and the chances of cure.
- Do not forget to mention social aspects of care and counseling especially for patients with serious chronic diseases like cancer and HIV.
- Finally, abide by the published peer reviewed guidelines of management approved by the authorities in your area.

MRCP(UK) Part II Exam (PACES)

As you already know, the long case has been cancelled and substituted by three stations: one for history taking (station 2), the second for communication skills and ethics (station 4), and the third is station 5 where focused history and examination is performed. All are observed performance stations.

History Taking

Is similar to that of the long case, but you are expected to exercise proper communication skills so that your interaction with the patient will be smooth and informative. Read the scenario in the provided letter from the referring source. Allow the patient to say any information not asked for.

(1) You should be able to elicit components of the history in a systematic and chronological order, and to identify the main problems with the patients and their effects on his daily physical and social activities.

(2) Periodically, agree with the patient on the main issues that have been discussed. Likewise, you are expected to discuss this with patient and agree on a management plan.

Communication and Ethics Station

(See the special section on communication skills.)

Much have been mentioned about this issue, and we have already referred to this earlier when we talked about history taking in general.

(1) A good doctor-patient relationship depends on the strength of the communication skills of the doctor. Patients prefer and like doctors who communication effectively with them. In fact, sound communication between doctor and patients and their

relatives will help in the management of even the complicated problems by ensuring patient adherence to treatment and mutual trust.

(2) First, greet, introduce yourself, relax the patient, and agree on the purpose of the interview. Confirm the patient name and ensure privacy and comfort of patient and relatives.

(3) Explore what the patient already knows about his problem. Example: 'Mr X, could you please tell me what you have been told or what you came to know about your problem?'

(4) Explain in simple language, avoid jargon(i.e., medical terms), the problem or information you want to convey to the patient and allow time for him/her to digest what you said throughout.

(5) Show sympathy and empathy and interact appropriately with the patient's feelings

(6) Answer the questions raised by the patient in a scientific and honest manner (e.g., prognosis, survival, outcome of chemotherapy, chances of cure, transmission of disease such as AIDS or hepatitis C or B).

(7) Allow the patient or his relatives enough time to express their concerns, and to discuss different aspects of management.

(8) Finally, agree on a management plan with the patient and his relatives including follow-up and support.

(9) Thank the patient before leaving.

The OSCE (PACES) Clinical Stations

The short case is the most reliable part of the examination that significantly tests the appropriateness of the techniques of physical examination to identify the correct physical signs.

1- The candidate should perform physical examination in a skillful, professional, and systematic manner.

2- He should be able to correctly identify abnormal findings, and report them in a systematic fashion.

3- He should give the appropriate interpretation of findings, so as to arrive at a sound diagnosis or differential diagnoses.

4- And then, outline a management plan that includes the necessary and relevant investigations, as well as general and specific treatments.

General Comments

- The time allowed to examine a patient as a short case is usually limited (six minutes in the MRCP PACES + four minutes discussion, and usually ten minutes in other boards).
- After washing hands, you should introduce yourself to the patient, greets him/her, and make her/him comfortable.
- Remember to be courteous and gentle to the patient.
- General look and position of patient:

    1. Sitting up: dyspnea, platypnoea
    2. Restless: shock, pain (MI)
    3. Holding chest (MI), dissection
    4. Lying away from light source (subarachnoid haemorrhage or meningitis)
    5. Leaning forward (pancreatitis, pericarditis)

- Observe the face first before hands, as most of the primary signs are there. Look for asymmetrical, shape, focal fits, and colourof the face.
- Then apply the general rule: see, feel, listen, and then talk. Or alternatively, talk after each step of examination if instructed by examiners.
- **It is more logical to start with examination of the face and neck**, where most of the primary signs exist, rather than the hands.(Facial look and shape: Cushing's, thyroid disease, and acromgaly. Also, Anaemia, jaundice, cyanosis, tongue and mucous memebranes, lymphadenopathy, neck swellings.)

Then proceed to hand examination looking for:

- Shape of fingers and toes: acromegaly, sickle cell disease
- Deformities: rheumatoid and osteoarthritis, gout and systemic sclerosis
- Skin: tightness, ulcers, and vasculitis (connective tissue diseases)
- Clubbing with or without cyanosis
- Stages of clubbing: one can just state whether clubbibg is early, established, or late drug stick, or grading as(1) nail bed fluctuation,(2) loss of angle between nailbed,(3) increased convexity of nail fold,(4) thickened distal,(5)hypertrophic osteoarthropathy
- Techniques, profile sign: first, look for the angle by holding the finer at the level of your eyes (early clubbing).
- View the fingers from a <u>dorsal</u> and <u>lateral</u> view. Note the width of terminal portion and compare with the proximal part.
- Look at the angle between the nail and skin.
- Inspect the <u>periungual skin.</u>
- Elicit <u>fluctuation</u> of the nail bed.
- Attempt to feel <u>the posterior edge</u> of nail.

In clubbing, there is widening of the AP and lateral diameter of terminal portion of fingers and toes, giving the appearance of clubbing.

- The <u>angle</u> between the nail and skin is greater than 180°.
- The periungual skin is stretched and shiny.
- There is fluctuation of the nail bed.
- One can feel the posterior edge of the nail.
- Look for fluctuation: softening and fluctuation of the nail bed.
- Then observe the curvature of the nails and confirm this by doing Schamroth sign.

Schamroth Sign (Window Sign)

In 1976, Schamroth reported a new clinical sign that incorporated two of the clinical features of clubbing. Normal fingers create a diamond-shaped window when the dorsal surfaces of terminal phalanges of similar fingers are opposed. In the clubbed finger, the diamond becomes obliterated because of the loss of the profile angle and the increase in the soft tissue at the cuticle. Since it's original description, this technique has become popular with physicians as a quick test to establish the presence of clubbing. The precision and accuracy of this sign, however, have not been formally tested. Also examine the toes for clubbing.

Remember the Causes of Clubbing

• **Respiratory diseases**

1. Suppurative lung diseases. This includes:

   • Bronchiectesis
   • Lung abscess
   • Empyema
   • Cystic fibrosis
   • Chronic TB or fungal cavities

2. Diffuse pulmonary diseases

   • Idiopathic pulmonary fibrosis
   • Asbestosis
   • Pulmonary arteriovenous malformations

3. Neoplastic conditions

   • Bronchial cancer
   • Malignant Mesothelioma
   • Bronchial fibroma
   • Metastatic osteogenic sarcoma

- Cardiovascular diseases

  - Cyanotic congenital heart diseases
  - Infective endocarditis
  - Arterial graft sepsis
  - Brachial arteriovenous fistula (unilateral)
  - Hemiplegic stroke (unilateral)

- Gastrointestinal diseases

  - Inflammatory bowel diseas
  - Malabsortion, like celiac disease
  - Hepatobiliary diseases and liver cirrhosis
  - Hepatocellular cancer

- Metabolic disorders: thyroid acropathy
- Familial: benign

  Then look for peripheral signs of liver disease: palmar erythema, Dupuytren'scontracture

Then Examination of Specific Systems

- **Is there any difficult system?**

The examination of any system can be difficult if the doctor has not acquired good beside skills. In the MRCP PACES, statistics say that the candidate scored lowest in the cardiovascular and central nervous system. This should not be taken for granted, but perhaps candidates should do more training with regard to the examination techniques at these systems, and in fact, all systems.

Examples of the clinical stations

1. The cardiovascular system

Expected cases in PACES and OSCE exams:

1. Aortic valve replacement
2. Mitral valve replacement
3. Aortic stenosis
4. Mitral regurgitation
5. Congenital heart disease (mostly VSD)
6. Aortic regurgitation
7. Marfan's with AR
8. Mixed aortic valve disease
9. Mitral stenosis
10. Pulmonary stenosis
11. Atrial Septal Defect
12. Coarctation of the aorta
13. Patent Ductus Arteriosus
14. Dextrocardia
15. Hypertrophic Obstructive Cardiomyopathy
16. Fallot's tetrology
17. Eisenmenger syndrome

Hints for CVS Examination

Important things to look for:

- I would measure the BP at this point (polycystic kidney, tranasplant) high arched palate/arachnodactyly/tall – Marfan's
- Pseudoxanthoma elasticum (mitral regurgitation)
- Sternotomy: look at legs for saphenous vein graft
- Short, webbed neck – Turner's: coarctation of the aorta
- Down's syndrome: ASD, VSD
- Hand-grip exercise – reduces AS/MS/HOCM, increased AR, MR, VSD

Things to mention at the end of your presentation:

1. Rhythm: regular /irregular
2. Presence of heart failure (LVF)
3. Presence of pulmonary hypertension or Cor pulmonale
4. Signs of infective endocarditis

Valvular Heart Diseases

a. Mitral valve disease (stenosis ± regurgitation), mitral valve replacement. This is more common in overseas exams, and less in Europe and North America.
b. Aortic valve diseases: aortic stenosis, regurgitation, or sclerosis. Also, aortic valve replacement. These are more common in Europe and North America.

The above can occur as a single lesion, but more often, in combinations.

- Young patients are more likely to have congenital, rheumatic heart disease, or infective endocarditis. While elderly patients have aortic valve disease due to ischaemic or degeneration.

Whatever the lesion is, the candidate should do the following:

(1) Look at the general examination such as the position in bed (at 45° or sitting up), pallor cyanosis, facial look, mitral faces, Marfa's).

Also, look for any O_2 apparatus, monitors, or any other equipments attached to the patient.

(2) Signs: inspect the chest and the precordium for surgical scars, central, and especially below the left breast. Also, thyroid enlargement or thyroid scars. Also look for peripheral signs such as peripheral cyanosis, clubbing, evidence of infective endocarditis, and nicotine staining.

(3) Look for sternotomy and thoracotomy scars.

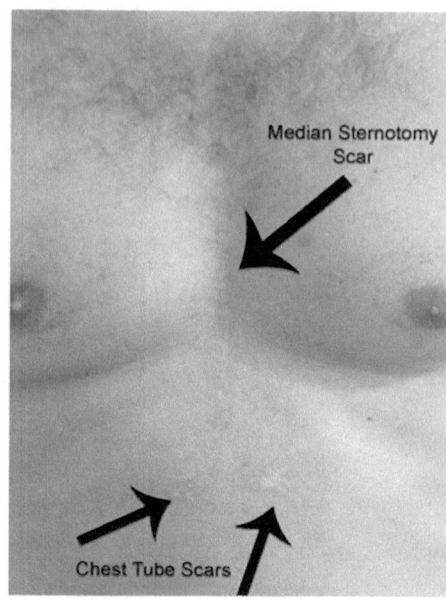

Fig 2: Inspection of the chest for cardiovascular purposes.

Remember to look and feel for devices such as defibrillators and pacemakers.

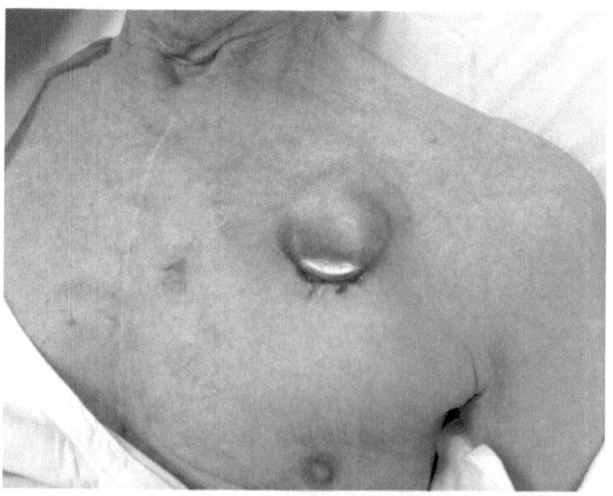

Fig3: Look and identify any devices.

The JVP and Carotids

This is very important and can be tricky. Follow the correct steps to identify the JVP and venous waves.

Distinguishing the internal jugular vein pulsations from the carotid artery:

Jugular Vein	Carotid Artery
No pulsations palpable	Palpable pulsations
Pulsations obliterated by pressure above the clavicle	Pulsations **not obliterated by pressure above the clavicle**
Level of pulse wave decreased on inspiration, increased on expiration.	No effects of respiration on pulse
Usually two pulsations per systole (x and y descents)	One pulsation per systole
Prominent descents	Descents **not prominent**
Pulsations sometimes more prominent with abdominal pressure	No effect of abdominal pressure on pulsations

(4) Examine the patient at 45° and sitting up to reveal very high JVP.

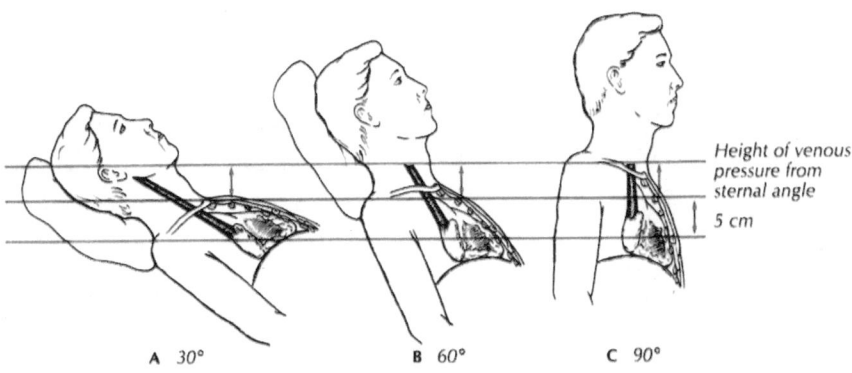

Fig 4: Position of JVP examination

Do the Hepatojugular Reflux (HJR) by applying gradual pressure below the right costal margin or mid abdomen if there is no tenderness, for up to a minute

Technique for Examining Hepatojugular Reflux (HJR)

Hepatojugular reflux is the distension of the neck veins precipitated by the maneuver of firm pressure over the upper abdomen. It is seen in tricuspid regurgitation, heart failure due to other non-valvular causes, and other conditions including constrictive pericarditis, cardia tamponade, and inferior vena cava obstruction. The HJR maneuver may be performed as follows:

- The patient is positioned supine with elevation of the head at forty-five degrees.
- Look at jugular pulsations during quiet respirations (baseline JVP).
- Apply gentle pressure (30–40 mm Hg) over the right upper quadrant or middle abdomen for at least ten seconds (some suggest to one minute).
- Repeat the JVP
- A positive response is defined by a sustained rise of more than 3 cm in JVPfor at least fifteen seconds after release of the hand. Patients must be coached to refrain from breath holding or a Valsalva-like maneuver during the procedure.

Note: Normal subjects will have a decrease in JVP with this maneuver since venous return to the heart will be reduced. The jugular venous pressure may transiently rise and then return to normal or decrease within ten seconds.

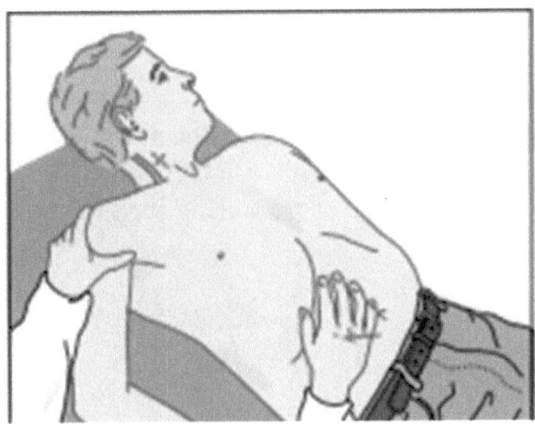

Fig5: The hepato-jugular reflux

The HJR helps:

- Confirm the waves are venous
- Makes the top of the venous column more obvious
- If present, confirms heart failure rather than venous obstruction
- Diagnose early right-sided failure

Identify the different waves, and whether they are arterial (Corrigan's sign) or venous (multiple, can be blocked by pressure at the root of the neck, and not palpable).

If the waves are venous, look for any prominent wave:

- Absent *A* wave: atrial fibrillation

    a. Giant *A* wave: pulmonary stenosis and pulmonary hypertension
    b. Cannon *A* wave: complete heart block, multifocal ventricular tachycardia
    c. Giant *V* waves: Tricuspid regurgitation

Remember:

The *A* wave will disappear if there is atrial fibrillation (and so is the presystolic accentuation of the mid-diastolic murmur of mitral stenosis).

- Perfect the technique of measuring the JVP.

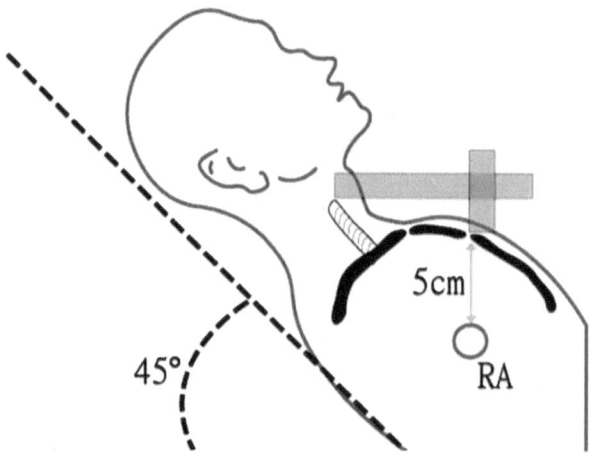

Fig6: Measurement of the JVP

- If the blood is pooled in the internal jugular and other veins without pulsations, consider the possibility of significant pericardial effusion or cardiac tamponade.

Examination of the Arterial Pulses
(Radial, brachial, carotid, femoral, and others)

Examine the pulses completely. Remember the following:

- Do the examination systematically. Ask patient if he has elbow or shoulder pain before raising the arm to look for collapsing pulse.
- Candidates sometimes get confused when performing the rhythm and character of the pulse.

- The rhythm of the pulse is either normal or irregular. Be sure about the type of irregularity, which can either be atrial fibrillation or extrasystoles (ventricular ectopics).
- If you are in doubt, exercise the patient. Ventricular ectopics will disappear, and also auscultate the apex to confirm AF.
- Look for pulse deficit.
- Do not forget to palpate the large arteries, especially the brachial and carotids (also auscultate), especially if you are looking for slow-rising pulse of aortic stenosis. The term anacrotic should be reserved when the slowrising pulse exhibits another impulse during the late phase of volume increase.
- Even some senior level registrars inadequately perform the technique of detecting the collapsing pulse. The usual technique is first asking the patient if he/she has any pain in the shoulder or elbow. Then use the mid-arm technique or palpate the brachial pulse, while briskly raising the arm.
- Pulsus bisferiens(combined aortic stenosis and regurgitation) may be difficult to detect if you are not sharp in identifying the double peaks (the early one represents the rapid rising and falling of collapsing pulse, while the late peak is that of the slow rising process of aortic stenosis).

It is most frequently caused by hemodynamically significant aortic regurgitation. Pulsus bisferiens is detected by examining the carotid upstroke. Two pulsations are detected in systole. The first is from the pressure increase related to left ventricular ejection. The second systolic pulsation is from either arterial recoil reflected from the periphery or actually early diastolic from the backflow of the regurgitant blood. The Valsalva maneuver or inhalation of amyl nitrate can precipitate pulsus bisferiens in some cases.

- Pulsus bisferiens can also be seen in Hypertrophic Obstructive Cardiomyopathy (HOCM), patent ductus arteriosus, arteriovenous fistulas, and normal hearts in a hyperdynamic state.

- **Collapsing pulse:**

  It is characterised by rapid upstroke (percussion wave) followed by rapid descent (collapse) of the pulse wave without dicrotic notch, which reflects low systemic vascular resistance.

- Rapid upstroke is due to the rapid ejection of greatly increased stroke volume.

  The rapid descent or collapsing character is due to:

  a) Diastolic 'run-off' (backflow) into the left ventricle
  b) Reflex vasodilation mediated by carotid baroreceptors secondary to large stroke volume.
  c) The rapid run-off to the periphery due to decreased systemic vascular resistance.

  Itis usually associated with capillary pulsation, which can be seen by applying pressure to the nails, or observed in the lips.

Pulsus Paradoxus (PP)

What is 'pulsus paradoxus'?

In fact, pulsus paradoxus is an exaggeration of normal physiology. It is an inspiratory fall in the systolic blood pressure greater than 10 mm Hg that occurs in the setting of:

1. Acute Pericardial tamponade (nearly 100 per cent of the time)
2. Asthma with FEV1 <0.7L.
3. Shock (approximately 50 per cent of the time)
4. PE (approximately 30 per cent of the time)

There are two ways by which PP can be detected. First, observe the drop in pulse volume during deep inspiration and then expiration. This can be difficult if the pulse rate is rapid.

Accurate Method to Detect Pusus Paradoxus

Inflate your BP cuff above the systolic. Lower the pressure very slowly until you hear your first Korotkoff sound. This first sound will only be heard during expiration. Continue to lower the cuff pressure to the highest value at which you hear Korotkoff sounds with each beat. This means that you are hearing sounds with inspiration and expiration. Find the difference between the two numbers, and this is the pulsus. Remember, a pulsus greater than 10 mm Hg is abnormal. If present, it indicates significant pericardial effusion, constrictive pericarditis, or severe bronchospasm.

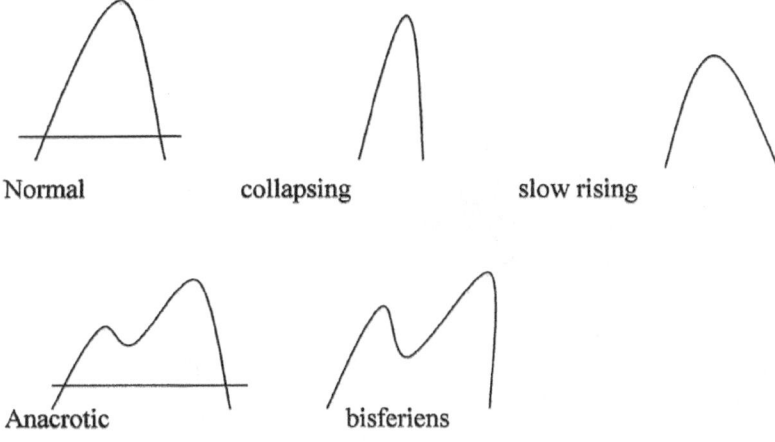

Normal	collapsing	slow rising
Anacrotic	bisferiens	

Fig7: Types of pulses

Completing the Pulse Examination

- First identify the right radual pulse using 2 to 3 fingers.
- Then examine the two radial pulses at the same time. If unequal, feel the two brachial arteries simultaneously. If the brachial pulsesare unequal, this suggests coarctation of the aorta beyond the right subclavian or compression of the subclavian or innominate by aneurysm or superior mediastinal mass. But if the brachial pulsesare equal, while the radialpulses are not, this strongly suggest an anatomical abnormality of the radial pulse.

- Aucultate the carotid pulses to make sure that they arepatent
- Do not forget to look for radio-femoral delay by palpating the femoral arteries.
- Also, palpate all peripheral arteries in the feet according to the correct anatomy.

Describing the Cardiac Impulse

- Avoid a description that fits a preconceived diagnosis.
- Feel meticulously with two fingers and see what your fingers are doing. Palpate the carotid at the same time to confirm the apex beat.

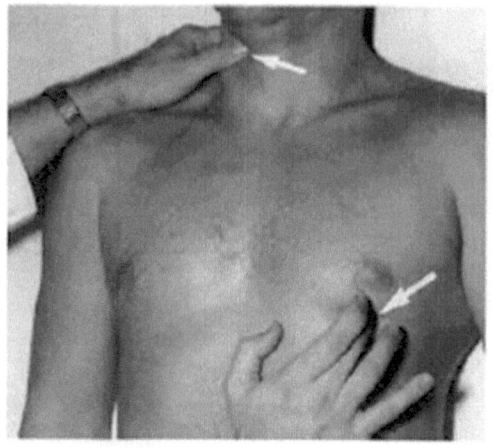

Fig8: Palpation for the apex beat

- The normal impulse is felt as a thrust in the usual position of the apex beat.
- If the apex beat is displaced, the impulse is likely to be abnormal in tow ways:

  1- Lifts your finger for a very brief moment. This means <u>unsustained heave</u>(some give the name forcible).
  2- Lifts your finger for a relatively prolonged time. This fits with the <u>sustained heave.</u>

The first type of impulse occurs usually in volume overload lesions, such as significant mitral and aortic regurgitation, while the second one (b) is caused by conditions that cause pressure overload (aortic stenosis, hypertension, some cases of hypertrophic obstructive cardiomyopathy HOCM where a double impulse can be felt).

- In case of right ventricular enlargement, the apex beat is usually in the normal place, and the impulse is described as tapping if the cause is pure <u>mitral stenosis</u> (a palpable loud S_1).

Other conditions, which may give a diffuse impulse, are:

1- Cor pulmonale
2- ASD
3- Patient ductus arteriosus

Undetectable Apex Beat and Impulse

- <u>First</u>, turn the patient to the left side. The apex beat will be seen and felt if it drops from <u>behind a rib.</u>
- <u>Second,</u> feel on the <u>right</u> side to look for dextrocardia. Then if the patient is obese or has emphysema, this will explain the situation.
- Other reasons for difficult to localsed apex beat are:

(1) Pericardial effusion
(2) Cardiac tamponade
(3) Obese subject
(4) Emphysema

Summary of Cardiac Impulse Examination

Abnormal Apex Beat Types

- Double impulse: Systole has two impulses.
  Possible diagnosis: hypertrophic cardiomyopathy

105

- Dyskinetic:
  Uncoordinated, easily palpable
  Possible interpretation: MI

- Hyperdynamic:
  Forceful, sustained apex beat
  Occurs in: AS, HTN

- Hyperkinetic:
  Coordinated, palpated beat is distributed over greater area.
  Occurs in: LV dilation

- Tapping apex:
  S1 sound is palpable.
  Occurs in: mitral stenosis

Auscultation

Bell OR Diaphragm

Low frequency sounds (S3, S4, MS rumble) are best heard with the bell applied lightly, whereas high frequency sounds (clicks, OS, MR) are best heard with the diaphragm. One can appreciate a spectrum of frequencies by applying differential pressure with the bell.

Braunwald suggests that the exam proceed from apex to Left Lower Sternal Border, then to left base, and from left base to right base. This follows the path of blood flow (inflow to outflow) and allows for physiological correlation with findings.

- When listening, identify the first (S_1), the second (S_2), or any other added sounds (S_3, S_4, summation gallop, opening snap, clicks, etc.).
- Be sure about the right place where each sound is heard, and do not make any confusion.

Patient Positions and Special Techniques for Auscultation	
Position	Use
Supine	General auscultation and most heart sounds
Sitting up and leaning Forward and holding exhalation	Aortic stenosis, aortic regurgitation, pericardial rubs
Left lateral decubitus	S3, S4, mitral stenosis (using bell of stethoscope)
Valsalva maneuver	increases intensity of mitral valve prolapse and hypertrophic cardiomyopathy, decreases intensity of aortic stenosis
squatting and standing	increases intensity of aortic stenosis, decreases intensity of outflow obstruction in hypertrophic cardiomyopathy

Types of murmurs:

A: Pansystolic (Holosystolic) murmur of mitral regurgitation

B. Late systolic murmur of mitral valve prolapse

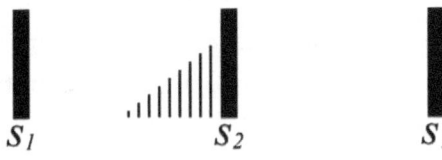

C. Early diastolic murmur of aortic regurgitation (soft blowing)

D. Rumbling Mid Diastolic diastolic murmur of mitral stenosis

E. Late diastolic (Presystolic) murmur of mitral stenos

Fig 9: Illustration of different murmurs

- A loud S_1 at the apex indicates mitral stenosis or tachycardia, and hence, will be associated with (rumbling diastloic murmur, opening snap, etc.). Therefore, if you heard any systolic murmur with a loud S_1, this means that the patient has <u>trivial mitral</u> regurgitation.
- On the other hand, if there is a significant pansystolic murmur at the apex with a third heart sound, severe mitral regurgitation is most likely present. Hence, there is <u>no reason to</u>say that the first heart sound (S_1) is loud. It got to be muffled.
- Also, if there is a loud mid-diastolic murmur with insignificant systolic murmur, together with a normal or low intensity S_1, one

can still diagnose mitral stenosis with valve calcification. In this case, the opening snap will not be found.

- The opening snap is high-pitched, heard by the diaphragm internal to the apex beat, and immediately follows the S_2. It indicates that MS is dominant and that the valve cusps are pliant and not calcified. The closer the OS to S_2, the severer is MS.

- Note that if there is Atrial Fibrillation (AF), the presystolic accentuation will be lost. So also is S_3.

- A loud P_2 (pulmonary component of S_2) indicated pulmonary hypertension. If associated with a loud S_1 and a rumbling murmur, one should talk about dominant MS and not MR. A soft P_2 indicates tetralogy of Fallot.

- Loud aortic component of S_2 is present in hypertension and also aortic aneurysm (described as ringing). A soft aortic component is a feature of aortic stenosis.

- The third heart sound (S_3) is low-pitched, localised to apex, and can be palpable. It is heard in mid-diastole and increase when the patient lies flat, also lost if there is AF.

- S_4 is a high-pitched sound heard later in diastole.

- Gallop rhythm: a gallop means the presence of three sounds (S_1, S_2, S_3 or S_4) with tachycardia.

- Maneuvers
  Remember to turn the patient to the left to make the murmur of MS louder. Also, to ask the patient to sit up and hold breath in full expiration to make the murmur of aortic regurgitation more audible.

Tricuspid Regurgitation (TR)

- Is diagnosed if there is large V wave in the JVP, pansystolic murmur, internal to apex and near lower left sternal border with a pulsatileliver.

- Causes: functional (congestive cardiac failure or cor pulmonale). (1) Organic: rheumatic heart disease with MS;(2) Rare: carcinoid syndrome, SLE, infective endocarditis.

Is it Aortic Stenosis?

Students and doctors are often confused when they hear a systolic murmur at the base (aortic area) that can be heard, sometimes in other areas including the apex.

Thus, to diagnose aortic stenosis, the following should be present:

a) A small volume pulse with a slow rising (anacrotic) character, best detected in big arteries such as carotid and brachial.

b) The apex beat is displaced downward and laterally with a <u>sustained heave.</u>

c) A soft aortic sound. If the aortic component of S_2 is normal or loud, then AS is in doubt (it could be sclerosis, hypertension, or aortic aneurysm).

d) A systolic thrill, which is usually not present in aortic sclerosis, but can be felt in aortic aneurysm.

e) An ejection mid-systolic (diamond-shaped) murmur that is conducted to the neck.

What indicates severe aortic stenosis?

Remembered as "ASH"

1. Angina and syncope and heart failure

Signs:

1. Slow rising pulse
2. Narrow pulse pressure
3. Thrill
4. Soft S2
5. S4
6. LVF

Remember:

Gallavardin phenomenon: part of the systolic murmur heard at the apex similar to MR

- DOES NOT RADIATE TO AXILLA!

Medical Treatment

Beware:

- Avoid nitrates.
- Be cautious with all antihypertensives, as they will either decrease cardiac output (beta blockers) or increase the gradient through vasodilatation.
- Cautious beta blockade or calcium channel blockade is appropriate.
- Digoxin and Furosemide can manage failure.
- ACEi are classically contraindicated, but are now viewed as safe in somepatients, although introduction should be done in hospital.
- If AS is accompanied with Aortic Regurgitation (AR), then the pulse will not be small in volume and pulsus bisferiens will be detected (two peak pulse), or collapsing if AR is dominant.
- Also, one will find some peripheral signs at AR such as capillary pulsations.
- In addition, causes vigorous pulsations at the root of the neck (Corrigan's sign). The pulse will be collapsing, and the femoral pulse is described as pistol shot.
- The typical murmur is a soft blowing early diastolic murmur best with the diaphragm, firmly pressed against the chest wall at the left lower sternal border during peak of expiration, while the patient is sitting up.
- This is one of the most difficult bruits to be heard, as it is soft and short.

- It is rarely accompanied with a thrill, and if so, the aetiology is most likely syphilic. Nowadays, in a young or middle-aged, it is usually due to rheumatic fever, Marfan's, or if there is aortic dissection.
- The murmur of AR may be accompanied with a rough mid-diastolic bruit called Austin Flint murmur (described in 1862 by the American physician Austin Flint). It is not true mitral stenosis as there is no loud S_1 or an opening snap, and its aetiology is most likely syphilic, which it is not common at present unless a patient has AIDS.

Peripheral Signs of Severe Aortic Regurgitation

- Capillary pulsations: **Quincke's sign**
- Waterhammer (collapsing) pulse and vigorous neck pulsations: **Corrigan's sign**
- Systolic head bobbing: **De Musset's sign**
- Systolic movement of uvula: **Muller's sign**
- Femoral retrograde bruit: **Duroziez's sign**
- Pistol shot femorals: **Traube's sign**
- BP lower limbs > BP upper limbs: **Hill's sign**
  - 20 mm Hg: **mild AR**
  - 40 mm Hg: **moderate AR**
  - 60 mmHg: **severe AR**

Additional Skeletal Signs Associated with Aortic Regurgitation

- ArgyllRobertson pupils: syphilis
- Marfan's syndrome: wide armspan, high arched palate
- Ank Spond: question mark posture
- Rheumatoid arthritis
- Pseudoxanthoma elasticum

What are the Causes of AR?

Acute:

- Infective endocarditis
- Dissection
- Trauma
- Prosthetic valve failure

Chronic:

- Rheumatic fever
- Endocarditis
- Syphilis
- Marfan's
- Ank spond
- Bicuspid aortic valve
- Hypertension
- Atherosclerosis

How do you grade the severity?

- Wide pulse pressure
- Soft second Heart Sound
- Duration of the early diastolic murmur
- Left ventricular failure
- Austin flint murmur: mid-diastolic at apex

If asked about the signs of severity of other valvular diseases, mention the following:

Severe Mitral Rugurgitation

- Small volume pulse
- Soft S1
- Third heart sound

- Enlarged left ventricle (specially failure)
- Pulmonary hypertension

Severe Mitral Stenosis

- Small pulse pressure
- Early opening snap
- Daistolic thrill
- Length of the diastolic murmur
- Pulmonary hypertension
- Atrial fibrillation

Severe Aortic Stenosis

- Plateau pulse (Parvus etardus)
- Single or paroxysmal splitting of A2
- Fourth heart sound
- Systolic thrill with long mid-systolic murmur
- Left ventricular failure

Clincal Manifestations of Severe Aortic Stenosis

Remembered as: ASH
These are: angina, syncope, andheart failure.

Gallivandin's phenomenon: Systmurmur of AS heardat the apex mimicking MR.

Septal Defects

Ventricular Septal Defects (VSD):

- Usually, the patient is either asymptomatic, has dyspnoea, or repeated respiratory infections.
- Lock for evidence of infective endocarditis.

- The pulse is usually normal.
- The auscultation findings depend on the stage of the disease.

1. Left to right shunt: by palpation, no evidence of right ventricular enlargement. The JVP is normal. Heart sounds are normal. The murmur is pansystolic, loudest at the lower left sternal edge, and does not radiate to the axilla. It is usually accompanied by a thrill.
2. Pulmonary hypertension

Feel for the left parasternal lift and palpable pulmonary sound (loud P2).

The thrill may distinguish, and then disappear when the flow decreases.

If there is reversed shunt, then cyanosis and clubbing may be seen (Eisenmenger complex).

Look for complications of VSD such as:

1- Infective endocarditis
2- Congestive cardiac failure
3- Eistenmenger complex
4- Heart block

Atrial Septal Defect (ASD)

Inspect the patient: Down's syndrome, fingers (hypoplastic thumb and extra phalanx, Holt-Oram syndrome)

- Is he cyanosed?If so, reverse shunt or Fallot's trilogy (ASD, pulmonary stenosis, and right ventricular enlargement).
- The apical impulse is normal or diffuse.
- Left parasternal lift

- The heart sounds wide fixed splitting of the second sound.
- The murmur is ejection systolic, loudest in the left second and third intercostal space.
- Remember, patients with ASD do not need prophylaxis for infective endocarditis unless it is accompanied by a vuvular lesion.
- ASD and acquired rheumatic mitral stenosis is called Lutembacher syndrome.

Points Often Missed

- Scars of previous surgery:

  - Valve surgery
  - Central scar of CABGE and open-heart surgery
  - Especially in females, thin scars below the breast

- Do not jump to murmurs. Describe heart sounds first:

  - Loud S_1 in the mitral area indicatesmitral stenosis.
  - Muffled S_1 suggests severe mitral regurgitation.
  - Variable S_1 suggests AF.

- If there is loud S_1, which suggests mitral stenosis (also occurs in tachycardia and hypertension), then look for opening snap

  → patient better on the Lt side, use usually the diaphragm, high-pitched, indicates free mobile cusps, no change with breathing.
  → If possible, comment on the distance of OS from the S_2 (0.04–0.10 seconds)
  → indicates the mobility of mitral valve cusp.
    Third heart sound S_3: if present, mitral regurg is predominant, and mitral stenosis trivial. Low-pitched.

- Do not mention contradicting statements that cannot be physiologically explained.

Examples:

- Loud S_1 in mitral regurgitation
- Loud aortic component of S_2 in aortic stenosis
- Tapping impulse with a pansystolic murmur at apex

Why do you need to examine the pharynx of the patient with a cardiac disease?

a. Looking to evidence of Marfan's (high arch palate): AR
b. Systolic pulsations of uvula (Muller's sign) in severe AR
c. Vasculitic lesions in infective endocarditis

Examination CVS Cases

Patient with shortness of breath

The patient was a 78-year-old female.

General: She was lying propped up in bed and had an oxygen mask on. Breathlessness in a patient with a cardiovascular problem was most likely to be due to heart failure(left ventricular failure or congestive failure).

Surroundings: no sputum pot, inhalers, or drugs.

The patient was of average height, but appeared to be thin. On examination of the head and face, the only physical sign of note was that the patient had flaring of the ala nasi in keeping with the suspicion of left ventricular failure. No pallor or cyanosis.

On examination of the hands, there was clubbing. In the context of the cardiovascular system, there are few causes of clubbing. In the absence of cyanosis, and in an elderly patient who was unlikely to have congenital heart disease (even after operative correction), the chances were the patient had infective endocarditis. There were no other peripheral stigmata of infective endocarditis.

The pulse rate was eighty beats per minute, completely irregular in rhythm and volume, indicating atrial fibrillation. Atrial fibrillation made the trainee suspects that the patient was likely to have mitral valve disease.

On examination of the neck, it was noted that the jugular venous pressure was elevated, and it was predominantly a systolic wave, which resulted in outward distension of the vein. This made you think it was a *V* wave due to tricuspid regurgitation. A *V*, wave in the context of atrial fibrillation and suspected mitral valve disease, would raise the suspicion that the patient had right ventricular involvement—secondary to the development of pulmonary hypertension as a consequence of mitral valve disease.

On examination of the praecordium, it was noted that there was a midline sternotomy scar. This raised the possibility of valve replacement or bypass grafting.

The apex beat was displaced to the sixth intercostal space in the anterior axillary line. It was thrusting (non-sustained heave) in nature (you turned the patient to the left side). This would indicate that the left ventricle was dilated, and there was diastolic overload. With the suspicion of mitral valve disease, already raised on account of atrial fibrillation, the most likely diagnosis at the moment was mitral regurgitation. There was left parasternal heave, and a palpable pulmonary second sound best felt at the pulmonary area. This was in keeping with the earlier suspicion that the patient had developed pulmonary hypertension as a consequence of mitral valve disease. On auscultation of the heart, the native first heart sound was replaced by a click, indicating a prosthetic mitral valve.

The second heart sound was complicated. At the pulmonary area, a loud native heart sound was heard in keeping with pulmonary hypertension. At the left sternal edge, a click was heard in keeping with a prosthetic aortic valve. (You sat the patient and auscultated during full expiration.)

In early diastole, an opening click was heard indicating that the mitral valve was a metal valve. There was a pansystolic murmur best heard at the mitral area, and this radiated to the axilla, confirming the earlier suspicion of mitral regurgitation. On auscultation of the lung bases, fine late inspiratory crepitations were heard in keeping with the earlier suspicion of left ventricular failure.

Diagnosis:

Prosthetic aortic and mitral valves
Mitral regurgitation
Atrial fibrillation, pulmonary hypertension, heart failure
Possibly infective endocarditis

Discussion:

What organisms would you suspect? How did the patient acquire endocarditis? How to investigate? If the cultures were negative, what organism would you suspect?

Management Plan

Information:

Bacteria-Free Infective Endocarditis

This rare typeof negative blood cultures occur in 2.5–31per cent of all patients of infective endocarditis, which often delays diagnosis and onset of treatment with profound impact on the clinical outcome

It occurs when there is an afebrile patient with strong clinical evidence of infective endocarditis but multiple blood cultures are negative. The manifestations of IE are due to prolonged stimulation of the immune system.

Usually, the causes for negative blood cultures are: (1) patients received antibiotics before blood cultures are taken due to systemic infection or suspected diagnosis of a bacterial infection; and (2) the suspected organism has limited proliferation or a special media is required.

Causes of culture negative endocarditis:

- Bartonella species
- Brucella species
- Chlamydia species
- Coxiella burnetii
- Legionella species
- Mycoplasma hominis
- Tropheryma whippelei
- Haemophilus aphrophilus
- Actinobacillus actinomycetemcomitans
- Cardiobacterium hominis
- Eikenella corrodens
- Kingella kingae (HACEK) organisms
- VRE (Vancomycin Resistant Enterococci)

Specific causative agents and the expected clinical features:

- Staph Aureus: This is a common and serious cause of IE.

It usually infects:

- IV drug users
- Patients with prothetic valves

- IV access linesif infected.
- Acute IE can occur due to staph bacteraemia.
- Conditions that cause impaired immunity is an added risk for IE. These includecancer, diabetes, corticosteroid use, IVDA, alcoholism, and renal failure.

  - Coaugulase negative S Aureus:

    This is responsible for about 30% of IE in prothetic valvaes & less than 5% of native valves. The disease is usually aubacute similar to strept viridians.

  - Streptococcus Viridans:

    This is a cause of about 60 per cent of subacute endocarditis and the clinical presentation is immune mediated.

- Nonenterococcal group D:

  - Is a cause of subacute IE
  - Source is large bowel with diseases such as ulcerative colitis, polyps, and bowel cancer
  - Usually sensitive to penicilline

- Pseudomonas aeruginosa:

  - Is the cause of acuteright-sided endocarditis in IV drug users.
  - Surgical treatmet is neededin most patients.

- HACEK organisms (refers to*Haemophilus aphrophilus, Actinobacillus actinomycetemcomitans, Cardiobacterium hominis, Eikenella corrodens, Kingella kingae*)

- Is the most common gram negative organisms causing subacute IE in about 5 per cent of patients.
- IE due to those organisms is complicated by congestive cardiac failure and massive arterial emboli.
- Appropriately treated with ampicillin and genamicin and possible surgery.

- Fungi:

- Usually causes subacute disease

- Candida albicans is main fungus infecting native and prothetic valve IE.
- In IV drug abusers, the most common fungus causing IE is Candida parapsilosis *or* Candida tropicalis.

- Bartonella:

  1. B Quintana causes IE in homeless people with bad hygene

- IE due to polymicrobials:

  2. The most common combination is pseudomonas and enterococciamong IV drug abusers.
  3. High surgical mortality

Prosthetic Valve Endocarditis

- Can be early that is within sixty days of valve surgery
- Complicated by severe congestive heart failure and symptoms of pericarditis or myocarditis.
- Embolic strokeis highest in the first three days of prosthetic valve.

Pacemaker Infective Endocarditis

- The clinical presentation depends on the part infected such as the generator pocket, intravascular leads, orepicardial leads.
- The type and source of infection

Fever is main presentation and may be the sole presentation in about a third of patients. It may occur acutely or subacutely. Late infections of the pocket may be due to erosion of the overlying skin without systemic involvement. Such erosions always indicate infection of the underlying device. Fver occurs in most patients.

Nosocomial Infective Endocarditis

Main manifestations are those of sepsis syndrome with hypotension, metabolic acidosis, and multiorgan damage.

The origin of sepsis is likely to be another organ like lungs (pneumonia) or the kidneys (pyelonephritis).

Central venous catheter is also a source of sepsis. The majority of IE (about 45 per cent) occurs in patients with prosthetic valves.

CVS Examination Exercise

Examine thispatient who complained of palpitations and fatigue.

The patient was a 65-year-old female lying comfortably propped up in bed. She was of average height and weight. Vital signs: comment on those as seen in the monitor.

There were no abnormalities seen on examination of her head. The face shape is normal and facial skin changes showed malar flush. No pallor or cyanosis or bleeding gums.

On examination of her hands, there were no abnormalities found in particular, no clubbing. Her pulse rate was 105 beats per minute, regularly irregular in rhythm, and normal volume andnormal character. All pulses were equal and synchronous.

On examination of her neck, the JVP was elevated to the angle of the jaw (10 cm). The dominant wave was an expansile systolic wave. The venous wave was under high pressure and palpable, suggestingtricuspid regurgitation and right heart failure due to pulmonary hypertension. As the patient did not appear to have significant pulmonary disease, the suspicion was that this was due to long-standing left ventricular dysfunction.

The trachea was in the midline. There were no abnormalities seen on inspection of the chest, the apex beat was palpable in the sixth left intercostal space lateral to the midclavicular line. Indicating that either the patient had systolic heart failure or dilatation of the ventricle due to diastolic overload caused by valvular regurgitation. There was no thrust or heave at the apex, and there was no parasternal heave. The first heart sound was soft, indicating that the mitral valve was not competent. The second sound was also soft. There was an opening snap heard best at the mitral area. This made think of stenosis of the mitral valve.

There was an ejection mid-systolic murmur radiating to the neck, suggesting that the patient has aortic stenosis. There was a pansystolic murmur at the left sternal edge, and this increased with inspiration in keeping with tricuspid regurgitation. This had been suspected earlier on the basis of the characteristics of the JVP.

There was also a pansystolic murmur at the mitral area. This increased in expiration, and radiated to the axilla in keeping with mitral regurgitation that had been suspected earlier on the basis of the soft first heart sound. There was a localised rumbling mid-diastolic murmur at the mitral area in keeping with mitral stenosis. There was no sacral or legoedema and the lung bases were clear.

Diagnosis:

- Mitral stenosis
- Mitral regurgitation
- Aortic stenosis
- Tricuspid regurgitation

Questions:

How would you know whether the mitral stenosis or regurgitation was dominant?

In this patient, one would suspect that mitral regurgitation was dominant. The pulse was of normal volume, the apex beat was displaced, and the first heart sound was soft.

How would you tell whether the aortic valve was severely narrowed?

The aortic stenosis was not severe in this patient because she had a normal volume pulse. There was no brachioradial delay, no thrill, no fourth heart sound. Can a mid-diastolic murmur occur in mitral regurgitation?

Causes of Mitral Regurgitation

Chronic MR:

- Functional MR from LV dilatation from any cause(IHD most common)
- Rheumatic fever
- Mitral valve prolapse
- Connective tissue disease
- SLE – Libman-Sachs endocarditis as in this patient

Rheum ArthritisandAnkylosingSpondylitis are other causes.

Congenital:

- Marfan's
- Ehlers-Danlos
- Pseudoxanthoma elasticum
- Endomyocardial fibrosis

Acute MR:

1. Trauma
2. Infective endocarditis
3. Papillary muscle or chordae tendineae rupture (post-MI)
4. Post-MS valvotomy

Criteria for severe MR and for repair:

- Heart failure
- S3 (rapid ventricular filling)
- Diminished EF
- Large LV dimensions

Acute MR may require surgery.

Is it Mitral Regurgitation or Prolapse?

Mitral valve prolapse:

- Undisplaced apex
- No heaves or thrills
- Normal heart sounds with a mid-systolic click and late systolic murmur in apex and and lower sterna border

Causes:

- Spontaneous(5 per cent more in women)
- Marfan's

- Ehlers-Danlos
- Polycystic kidney disease (PCKD)
- SLE
- ASD

Complications:

- MR
- Arrhythmias
- Atypical chest pain
- TIA/CVA, especially if associated with ASD or PFO

Infective endocarditis

- Sudden cardiac death

Other CVS Problems

Hypertrophic Obstructive Cardiomyopathy (HOCM), usually young patient who present with syncope.

History:

- Asymptomatic
- Dyspnoea on exertion
- Chest pain
- Syncope
- Dizziness and palpitations
- FH history of sudden death

Examination:

- Bifid carotid pulse
- JVP *A* wave
- Double apical impulse
- PSM at apex from MR

- ESM at LSE increased by Valsalva and standing, decreased by squatting

Investigations:

- Echo – gradient across valve, systolic anterior motion of mitral valve leaflet, asymmetric hypertrophy, MR
- ECG – LVH, AF, LAD, RBBB
- CXR – may show atrial enlargement

Complications of HCM:

- Sudden death
- AF/arrhythmias
- Infective endocarditis
- Systemic embolism

Treatment:

- Education with genetic counselling
- Prevention of dysrhythmias with anti-arrhythmics or with pacing/ICD
- Medical management of LV dysfunction with betablocker, diuretics
- Septal ablation or myomectomy

Pulmonary Stenosis

History:

- Maternal rubella
- Asymptomatic
- Dyspnoea or fatigue

Examination:

- Round facies
- Normal pulse
- *A* wave on JVP
- L parasternal heave
- Ejection click
- Soft P2 with split S2
- ESM in LUSE on inspiration radiating to left shoulder

If cyanosis and clubbing, think ToFallot.

Causes:

- Congenital(Noonan, Williams)
- Rheumatic
- Carcinoid

Cardiac Problems in Systemic Diseases

Condition	Appearance	Associated cardiac abnormalities
Marfan Syndrome	Tallandlong extremities	Aortic root dilatation Mitral valve prolapse
Acromegaly	Large stature, coarse facial features, spade hands	Cardiac hypertrophy, cardiomyopathy, resistant heart failure, Hypertension
Turner syndrome	Web neck Hypertelorism Short stature	Aortic coarctation Bicuspidaortic valve Pulmonary hypertension Cor pulmonale

Pickwickian syndrome Obesity hypoventilation syndrome	Severe obesity Somnolence	Hypertrophiccardiomyopathy, Cardiomyopathy
Friedreich ataxia	Lurching GaitHammertoePes cavusLV dilatation Pseudohypertrophy of calves	Aortic regurgitation Heart block (rare)
Duchenne type muscular dystrophy	Straight back syndrome	Right-sided congestive heart failure Prosthetic valve dysfunction (hemolysis)
Ankylosing spondylitis	Stiff ('poker') spine Yellow skin or sclera	Pulmonary hypertention Secondary cardiomyopathy Hypertrophic obstructive cardiomyopathy Pulmonary stenosis Pulmonary arteriovenous fistula
Jaundice		
Sickle cell anemia	Cutaneous ulcers Painful 'crises'	
Lentigines (LEOPARD syndrome*)	Brown skin macules that do not increase with sunlight	Cathecolamine-induced secondary dilated cardiomyopathy
Hereditary hemorrhagic telangiectasia (Osler-Weber-Rendu disease)	Small capillary hemangiomas on face or mouth, with or without cyanosis	
Pheochromocytoma	Pale, diaphoretic skin Neurofibromatosis – café-au-lait spots	

*LEOPARD syndrome: Lentigines, Electrocardiographic changes, Ocular hypertelorism, pulmonary stenosis, abnormal genitalia, retardation of growth, deafness

Condition	Appearance	Associated cardiac abnormalities
Lupus	Butterfly rash on face Raynaud phenomeneon-hands Livedo reticulars Cutaneous nodules	Verrucous endocarditis Myocarditis Pericarditis
Sarcoidosis	Erythema nodosum Angiofibromas	Secondary cardiomyopathy Heart block
Tuberous sclerosis adenoma sebaceum	Coarse, dry skin Thinning of lateral eyebrows	Rhabdomyoma
Myxedema	Hoarseness of voice Cyanosis and clubbing of distal extremities Differential cyanosis and clubbing Rudimentary or	Pericardial effusion Left ventricular dysfunction, Any of the lesions that cause Eisenmenger syndrome
Holt-Oram syndrome	absent thumb, Mental, retardation,	Reversed shunt through patent ductus arteriosus
Down syndrome	small mouth,	Atrial septal defect
Scleroderma	Tight, shiny skin of fingers with contraction Characteristic taut mouth and facies	Pulmonary hypertension Myocardial, pericardial or endocardial disease

	Typical hand deformity Subcutaneous nodules	Myocardial, pericardial or endocardial disease (often subclinical)
		Mitral valve prolapse
	Pectus excavatum Straight back	
Rheumatoid arthritis	Reddish cyanosis of face Periodic flushing	Pseudocardiomegaly
Thoracic bony abnormality		Right-sided cardiac valve stenosis or regurgitation
Carcinoid syndrome		

Practical Exercise

A 33-year-old lady presented with chest discomfort.

Start off by having a general look at the patient and her surroundings. You noticea yellow booklet by the bedside, which indicates that the patient was on warfarin. Thus, you conclude that the patient has either an arrhythmia such as atrial fibrillation or a metallicheart valve.

Inform the examiner that this young female is comfortably sitting in bed and looks rather slim. She looks tanned, meaning that she was recently arrived from a holiday in atropical country.

You logically started with head, face, and neck examination (not hands) looking for the primary physical signs. She has normal facial look (not plethoric or cushinoid) without pallor, jaundice, or cyanosis. Examination of the neck did not reveal any lymphadenopathy or thyroid swelling. The JVP was not raised. Examination of the hand showedher pulse rate to be eighty-four beats per minute, of good volume and charachter, and regular. There is no radioradial or radio-femoral delay.

As she was in sinus rhythm, you conclude that the indication for warfarin was a prosthetic valve rather than atrial fibrillation. Examination of the chest showed a midline sternotomy scar, which strongly supports the fact that she underwent cardiac surgery. When examining the back of the chest, aleft-sided thoracotomy scar, suggesting suregery for repair of coarctation of the aorta, was detected. The apex beat waslocated in the fifth IC space medial to the midclavicular line with a normal impulse charcter.

There was no left parasternal lift or palpable pulmomary componenet of the S2. The first heart sound was normal, but there was a metallic click replacing the second heart sound. As well, after listening carefully, there was an opening clikc at the left sterna border during systole. All these support that the patienthas a metallic valve.

Listening at the aortic area did no reveal any murmurs even when asking the patient to sit up, breathe deeply, and hold breath, looking for aortic diastolic murmurs. Then, when auscultating the lung bases, no rales were heard. Abdominal examinations show no organomegaly or ascites.

Examination of legs and sacrumshowed no oedemaand the peripheral pulses were intact.

What is the full diagnosis?This patient has a prosthetic aortic valve replacement of a bicubid aortic valve associated with coarctation of the aorta. The prosthetic valve is wellfunctioning, as there is no evidence of leak. She is in sinus rhythm and not in heart failure.

Brief Cardiac Summary Notes

Valvular heart disease

Aortic Stenosis:

- The pulse is slow rising, BP with narrow pulse pressure
- The apex beat is laterally displaced, and the impulse has a sustained heaving character, (pressure overload).
- S2 is soft
- There is a mid-systolic ejection systolic murmur radiating to the neck. Might be accompanied with mild aortic regurgitation

Aortic Regurgitation:

- Pulse: large volume collapsing
- BP: wide pulse pressure
- Displaced apex with non-sustained heaving impulse (volume overload)
- Earlyblowingdiastolic murmur at first aortic, and second aortic area at the left lower sternal border. Best heard when the patient site up and holds breathing in deep expiration.
- Corrigan's sign at neck, duroziez, and other peripheral signs.
- Causes: infective endocarditis, syphilis, rheumatic. Others: ankylosing spondylitis, Marfan's

Mitral valve disease:

- **Mitral stenosis**: malar flush, small volumepulse with possible AF, nondisplaced tapping impulse, while displaced non-sustained impulse in mitral regurgitation (volume overload).
- Loud S1, possible opening snap in MS while soft S1 in MR.
- Murmurs: rumbling mid-diastolic murmur in MS with possible presystolic murmur. In MR: holosystolic murmur radiating to axilla.
- Signs of pulmonary hypertension in MS

- Causes(MS): rheumatic, while MR is caused by rheumatic, infective endocarditis, mitral valve prolapse, infective endocarditis, connective tissue disease, and degenerative

Tricuspid Regurgitation:

- Aised JVP with large *V* wave
- Signs of pulmonary hypertension
- Pulsatile hepatomegaly
- Pansystolic murmur at the lower left sternal area increased with inspiration

Quick reminders:

AR: look for infective endocarditis, Marfan's, ankylosing spondylitis, ArgyllRobertson pupils (syphilis)

Notes:

- AS: Valve replacement indicated if symptoms or when gradient > 50mmHg, valve area < 0.6
- AS: Beta blockers slow rate of rise of systolic pressure, avoid vasodilators, ACE inhibitors
- AS: Impossible to have mod/severe with normal S2
- Austin flint mid-diastolic murmur occurs in severe AR due to regurgitant jet interfering with mitral valve
- AR: Surgery when either (i) EF<50%, (ii) decreased exercise capacity, (iii) LV end systolic 55mm, (iv) aortic root 50mm
- AR: Long-term nifedipine reduces/delays the need for valve replacement in asymptomatic patients with severe AR, not recommended for mild AR as good outcome with no therapy
- Acute MR due to papillary rupture in MI, endocarditis
- MR indications surgery: Either (i) mod/severe symptoms despite medical therapy, or (ii) EF<60% / LV end systolic dimension >45mm in absence of symptoms
- MS: normal valve area 4–6 cm2 tight stenosis < 1

- MS indications intervention: symptoms/puloedema/ haemoptysis/pul hypertension
- MS indications percutaneous valvotomy: mobile valve, not calcified on TOE
- Graham steele: EDM of pulmonary regurg with pulmonary hypertension
- Tissue aortic valve can have just loud S2
- Metallic S1 = mech mitral, S2 = mech aortic, both = either double valve or ball and cage

2. The Gastrointestinal System

General Considerations: for examination

- The patient should have an empty bladder.
- The patient should be lying supine on one pillow, and appropriately exposed from nipple to symphesis pubis (then to mid thigh if genetalis are examined).
- The examination room must be quiet to perform adequate auscultation and percussion.
- Watch the patient's face for signs of discomfort during the examination.
- Use the appropriate terminology to locate your findings:

  o Right Upper Quadrant (RUQ)
  o Right Lower Quadrant (RLQ)
  o Left Upper Quadrant (LUQ)
  o Left Lower Quadrant (LLQ)
  o Midline

- Epigastric
- Periumbilical
- Suprapubic
- Disorders in the chest will often manifest with abdominal symptoms. It is always wise to examine the chest when evaluating an abdominal complaint.

- Consider the inguinal/rectal examination in males. Consider the pelvic/rectal examination in females.

Abdomen stations and short cases

1. Splenomegaly (lymphoma, leukemia, portal hypertension, sherocytosis)
2. Patient on dialysis (uraemic changes, fistula,)
3. Polycystic kidney disease (hypertension, palpable kidneys, fidtula)
4. Renal transplant (side effects of immunosuppressants, palpable mass R L Q)
5. Chronic liver disease (jaundice, parotid enlargement, ascites, dilated veins.)
6. Hepatosplenomegaly (many causes)
7. Ascites (liver, kidney, cardiac disease, malignancy...)
8. Unilateral enlarged kidney (nephrectomy scar, uraemia, ? AV fistula)
9. Crohn's disease (surgery scars, skin, anaemia...)
10. Primary Biliary Cirrhosis (jaundice, xsanselasma, portal hypertension...)
11. Liver transplant (typical Mercedes scar, features of liver disease.)
12. Jaundice (many causes, liver disease, haemolysis.)

Chronic Liver Disease

History:

Fatigue, weight loss, jaundice, increasing abdominal girth, mental state changes

Cause:

- Alcohol
- Hep B or C (transfusion, sex, IVDU)

- Medications: (amiodarone, methyldopa, methotrexate)
- Diabetes? Tan? Haemochromatosis
- Chorea? Wilson's
- Emphysema? Alpha1AT
- FH of liver disease

Signs:

Face: Congested (alcohol), jaundice, xsanselama, parotid enlargement, bleeing)

- Peripheral signs: hands
- Clubbing
- Leukonychia
- Dupuytren's contracture
- Palmar erythema
- Spider angiomata (naevi)
- Tattoos
- Hepatic flap
- Pallor
- Scratch marks
- Pigmentation
- Loss of hair
- Gynaecomastia
- Splenomegaly and ascites
- Hepatomegaly(asses liver span)

Causes:

- Alcohol is the most common in the developed countries
- Viral hepatitis B and C is next most common, and the most common worldwide
- Autoimmune (PBC, PSC, AIH 1 and 2)
- Metabolic(Haemochromatosis, Wilson's, alpha1AT)
- Drugs(Methotrexate, Amiodarone)
- Cryptogenic

Investigations:

- Initial workup: blood for CBC and liver enzymes and function
- Blood tests for causes (immunoglobulins, ceruloplasmin, hepatitis serology…)
- US Abdo(size, thrombosis, hepatoma, portal hypertension)
- Alpha Fetoprotein

Complications:

- From hepatic dysfunction:

  o Encephalopathy
  o Coagulopathy
  o Hepato Cellular Cancer
  o Nutritional problems

- From Portal hypertension:

  o Ascites
  o SBP
  o Variceal bleeding
  o Hepatorenal and hepatopulmonary syndromes

Child Pugh

Prognosis from chronic liver Dx (Synthetic and metabolic function):

- Bilirubin
- Albumin
- INR
- Ascites
- Encephalopathy

MELD score (equivalent):

- Bilirubin
- INR
- Creatinine

Primary Biliary Cirrhosis

History:

- Fatigue
- Pruritis
- Jaundice
- Xanthomata
- CLD complications

Examination:

- Middle-aged woman
- Icteric with skin pigmentation
- Xanthelasma/xanthomata
- Excoriations
- Hepato/hepatosplenomegaly

What is PBC?

An autoimmune disease of the liver with progressive destructions of small bile canaliculi leading to cholestasis, fibrosis, and cirrhosis(90 per cent female). It is associatedwith a range of other autoimmune conditions including Sjogren's, RA, SS, Hashimoto's, and Coeliacs. It may be triggered by exposure to a pathogen.

Idiopathic Haemochromatosis

Examination:

- Slate grey pigmentation
- Decreased body hair
- Gynaecomastia
- Testicular atrophy
- Hepatosplenomegaly with CLD

Other signs:

- Cardiomyopathy +/- arrhythmias
- Diabetes
- Arthropathy
- Addison's disease
- Hypothyroidsim
- Hypogonadism

What haemochromatosis?

Hereditary haemochromatosis is a genetic, autosomal recessive iron storage disordercharacterised by a state of iron overload, which causes damage to a range of organs.

Classical (type 1) HH caused by HFE gene mutation; however, other forms involvinghepcidin, transferrin, and ferroportin mutations are known. Secondaryhaemochromatosis can occur from chronic haemolysis, repeated blood transfusion, orexcess iron intake (rarely).

Investigations:

- Haematinics(ferreting and transferring saturations)
- HFE genetic analysis
- MRI
- Liver Biobsy

Treatment:

- Venesection (aim for ferritin below a certain level)
- Deferoxamine – chelation agent – infusion
- Exjade – chelation agent – oral
- Genetic counselling and family testing

Liver Cirrhosis(Hepatitis B and C, Alcohol, Bilharzia) and Fibrosis

- Ascites + Shnuker liver + splenomegaly (portal hypertension)

    - As before: introduce yourself, greet, explain purpose, and the start with inspection. Call patient bythe first or favourable name.
    - Make sure position of the patient is correct(i.e., flat with onepillow, or two if distressed).
    - Exposure: above nipples to mid-thigh. Butcover genitalia and examine when appropriate.

Look for:

(1) roundedface (moon: cushing's), plethoric (cirrhosis, polythycemia). Shape: wasted (malignancy, malabsorption)
(2) Eyes: pallor

   Jaundice
   Kayser-Fleischer ring
   Xanselasma, arcus

(3) Parotid enlargement (alcoholic cirrhosis), Sjogren's syndrome
(4) Cheeks: visible capillaries

   Paper money sign, which is a complex of dilated capillaries common inalcoholic cirrhosis

(5) Mouth: hygiene, tongue (pale and smooth indicates iron deficiency anaemia. red, glossitis in megaloplastic anaemia)

Angular stomatitis
Telangiectasia: Hereditary
Haemorrhagic Telangectasia, candida

(6) Petichae in palate and mucous membranes. Then examine neck for enlargement lymph nodes, especially supraclavicular (TB lymphoma malignancy)

Virchow's Node (troussaeu sign):

(7) Single lymph node enlargement
(8) Left <u>Supraclavicular Lymphadenopathy</u>
(9) Behind clavicular head of sternocleidomastoid
(10) Difficult to localise due to deep location

Technique pointers:

(11) Palpate behind anterior head sternocleidomastoid
(12) Perform exam with patient using Valsalva maneuver

Interpretation:

(13) Indicates metastasis from abdominal malignancy

References:

(14) Degowin (1987) Diagnostic Exam, Macmillan, p. 222
(15) Wilson (1991) Harrison's Internal Medicine, p. 354–6

Examine hands (better after face)

- Pallor
- Palmar erythema

- Pigmentation (haemochromatosis), tylosis palmaris (GI malignancy)
- Dupuytren's contractures: feel palm and observe deformity of fingers (alcohoic cirrhosis)
- Clubbing, koilonychia
- Asterixis flapping tremor: indicates hepatic encephalopathy

Examine upper chest for:

- Gynaecomastia and spider angiomas (naevi)
- Dilated veins (venous obstruction)

Then examine abdomen

- Observed general contour from foot of the bed (observe flanks) and shape(normal, scaploid, or distended).
- Visible peristalsis
- Petechiae
- Look for bleeding around the umbilicus(Cullen's sign) and on the flanks(Grey Turner signs). Cullen's and GreyTurner's signs have been described with intra-abdominal haemorrhage most commonly associated with pancreatitis. Rare associations include portal hypertension with coagulopathy, ectopic pregnancy, malignant disease (liver, abdominal metastasis), perforated duodenal ulcer, liver abscess, and splenic rupture.

B   **A**

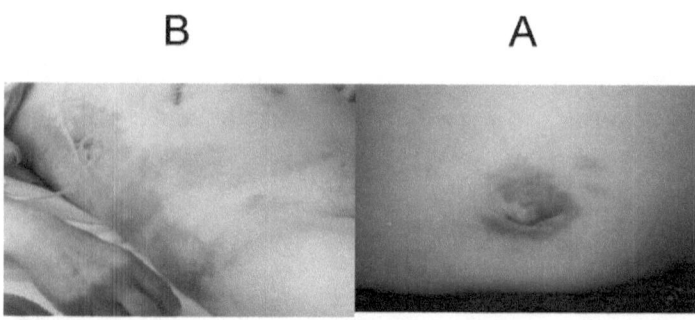

Fig 10: signs of acute pancreatitis: (A: Cullen's, B. Grey Turner's)

- Umbilicus (caput medusae), umbilical hernia bulging

  Caput medusae means 'Medusa's head'. Medusa is a Greek mythological figure, ruler of the Gorgons. In her mortal form, she had a head covered with snakes and looking into her eyes would cause an observer to turn into a stone statue.

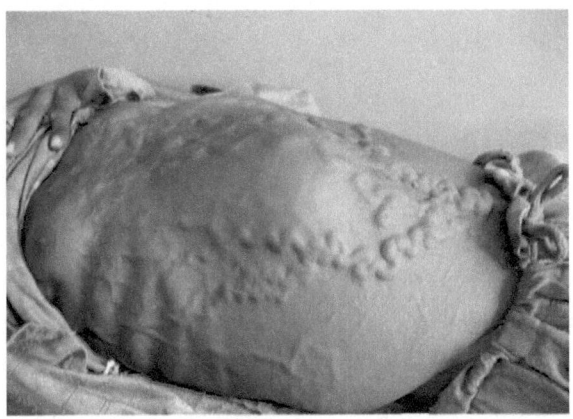

Fig 11: Huge Dilated Veins: Portal hypertension/
Inferior venecava obstruction.

If flow in all veins above and below the umbilicus is upwards, this indicates inferior vena cava obstruction, and if flow of all veins is downwards, the this is superior vena cava obstruction. But if flow of the veins below the umbilicus is downwards and upwards in those above the umbilicus, then this is portal hypertension

- Genitalia: female hair distribution in males (also axillae). Testicular atrophy (sigs of cirrhosis)

(1) Palpate

   a. Superficially forareas of tenderness and superficial masses. observe face of the patient while you are palpating.

(2) Deep palpation

Hints:

b. Do not use tips of fingers
c. Do not inflict pain (observe facial expressions).
d. Ask patient to breath in and out and move your palpating palm upwards during inspiration.
e. Remember the ABCs of palpating for the spleen and liver: start in the right iliac fossa, and go inch by inch as stated above.

Detection of Enlargrd Spleen

Palpate, as mentioned above.

f. If you felt a mass extending downwards from the left hypochondrium, make sure whether it is the spleen or not. If it is the spleen:

    a. Cannot push your hand above the mass and below the left hypochondrium
    b. Look for the notch
    c. Moves downwards during inspiration
    d. Dull to percussion

g. Measure the size of the spleen in the midclavicular line.
h. If the spleen is not palpable, turn the patient on right side and palpate again. Also, percuss the last two intercostal spaces and traube area
i. If the spleen is not palpable, then percuss the Traube's area. The boundaries are the anterior axillary line, the costal margin, and the sixth rib. Dullnes indicates splenic enlargement.

Castell's method:

j. Percussin of the eighth or ninth space in the anterior axillary line during full inspiration. If dullness is found, then the spleen is enlarged.

    k. You may auscultate over the spleen for presence of splenic infarct in long-standing splenic enlargement.

Nexon's method(not usually needed):

1. Percuss in a line perpendicular to the left costal margin in the right lateral decubitus position.
2. If the percussion extends upwards >8 c, then this indicates possible splenic enlargement.

Remember the differential diagnosis of a huge splenomegaly:

- Infections

    (1) Visceral leishmaniasis (kalazar)
    (2) Portal hypertension (bilharziahepatitis)
    (3) Chronic brucellosis
    (4) Tropical splenomegaly syndrome (hyperreactive malarial splenomegaly)
    (5) Some cases of disseminated tuberculosis

    Metabolic

    (6) Glycogen storage diseases (Gaucher's)
    (7) Amyloidosis

    Some haemolytic anaemia:

    1. Thalassaemia (sickle cell thalasaemia)

    1- Congenital splenocytosis

- Myeloproliferative diseases:

    2- Chronic myeloid leukaemia
    3- Myelofibrosis
    4- Hairy cell leukaemia
    5- Polycythemia rubra vera

- <u>Lymphomas:</u>

  6- Non-Hodgkin's
  7- Primary lymphoma of the spleen

Examining for Liver Enlargement

a. Palpation as for the spleen from RIF
b. Find the lower edge and mark it
c. Then percuss the chest from above (2d right space) to define the upper border of the liver (usually fifth to sixth space).
d. When you detect dullness, ask the patient to take deep breath and percuss during inspiration (dullness should diminish) and expiration. <u>This is called tidal percussion.</u> This is to make sure that the dullness is not due to respiratory disease (e.g., pleural effusion or pleural thickening causes fixed dullness).
e. Percuss from below and measure the liver span.
f. Liver span is 9–12cm according to gender (less in women) and height (less in short people).
g. If there is ascites, do not percuss from below. Use dipping(ballottment) to detect the lower border of the liver.

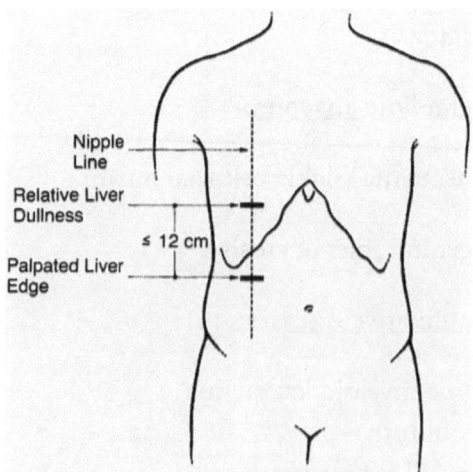

Fig 12: Estimation of liver span

Normal liver spans:

Hight (cm)	men	women
150	8.25	6
158	9	6.75
165	9.75	7.5
180	11	8
188	11.8	9.7

h. Define the texture, surface, and if nodular, find whether it is fine or coarse nodularity.
i. Find out whether there is a single mass and find its location in the liver.
j. Report on tenderness, pulsation (in tricuspid regurgitation or transmitted from below).
k. Listen for bruit (systolic murmur) or venous hum (continuous flow in the collaterals of portal hypertension).

Remember:

a. The liver edge can be felt below the right hypochondrium even if the liver span is normal. This occurs in:

- Thin people
- If the liver is pushed down by emphysema or any chest pathology.
- If there is a Riedel's lobe.

b. Also, the liver can be normal in size, but pushed upwards by ascites or tumour or large abscess. Here, the upper border will be displaced upwards (fourth or third space).
c. Remember to feel the abdomen for intra-abdominal lymph nodes, palpable kidneys, and aortic aneurysm.
d. Also, listen above the umbilicus on either side for the bruit of renal artery stenosis.

e. Abdominal examination is not complete without examining the genitalia for hair distribution, and size of testicles in men (ask permission to expose the patient and stop if denied).

Causes of Hepatomegaly

Hepatomegaly

Examination:

1. Size, tenderness, surface, span, bruit
2. Is there an enlarged spleen?
3. Is there any ascites?
4. Are there signs of chronic liver disease?
5. Are there any Lymph nodes?
6. Is there an indication of a primary tumour?
7. Is there a raised JVP/signs of right heart failure?

Top 3

1. Liver fibrosis/chronic hepatitis/cirrhosis, most likely alcohol (signs of CLD)
2. Primary or secondaryMitotic lesion (evidence of primary, clubbing, cachexia, irregularandnodular liver)
3. Congestive heart failure (tender): (Raised JVP, S3, pulm heave, ankle oedema, TR, giant *V* waves, pulsatile, and tenderliver)

Others:

- **Infections**: Hepatitis (B and C), EBV
- **Neoplasms of the liver**: benign or malignant, primary or secondary
- **Lymphoproliferative Dx** (LNs)
- **Primary biliary cirrhosis** (middle-aged female with jaundice, scratchmarks, and xanthelasma)

- **Haemochromatosis** (male, grey pigmentation, resistant DM)
- **Sarcoidosis** (lupus pernio, MZ crackles, skin signs)
- Reidel's lobe
- Emphysema with apparent hepatomegaly
- **Amyloidosis** (RA or other chronic Dx)
- Budd-Chiari (icterus, ascites, tender hepatomegaly)
- Hydatid cyst (Welsh)
- Amoebic abscess (drain sites, ethnic origin)

Hard and nobbly liver?

- Malignancy
- Polycystic liver Dx (feel for kidneys)
- Macronodular cirrhosis (rare)
- Hydatid cysts
- Syphilitic gummata

Causes of Large and Tender liver

- Acute hepatitis
- Congestive heart failure
- Liver abcess(amoebic or pyogenic)
- Acute Malaria
- Acute leukemia

Other short cases:

Patient with distended abdomen(Ascites)

a. Such patients are usually having ascites or huge abdominal masses (large spleens, livers, para-aortic lymph nodes, pancreatic cysts; or in females, gynecological masses).
b. Start as mentioned before (general examination).
c. Then observe the pattern of distension. Comment on the flanks.

d. If it is generalised distension including the flanks meanshuge ascites.

e. If there is central distension with normal flanks, this is suggestive ofencysted fluid, ovarian cyst (in women). Or may be a solid mass or masses.

f. After inspection, do superficial palpation and then confirm ascites by:

(1) Fluid wave sign (fluid thrill) if moderate or larger distension.
(2) If mild distension, perform shifting dullness(never do if there is huge ascites).

When performing shifting dullness:

1. While patient supine, percuss from epigastrium downwards till you confirm resonance due to bowel(fingers transverse).

Technique 1(easier):

With the patient lying flat, start percussing from the midline away from you. If the percussion note becomes dull, hold your index finger in that position and your thumbin the resonant area. Then ask the patient to roll towards you and wait for about thirtyseconds. Thenpercuss over this area, and if the note changed to resonanace, then shifing dullness is positive indicating ascites.

Technique 2:

Percuss from the midline on both sides (fingers longitudinal) unless there is a mass. When dullness is detected, mark the point with a skin marker.

Then turn the patient away form you (on his left side, as it iseasier to repercuss than turning towards you). Wait for about thirty seconds.

(1) Positive test: When ascites is present, the area of dullness will shift upwards towards the umbilicus, and the area of tympani will move upwards towards you.

The shift in zone of tympany with position change will usually be at least 3 cm when ascites is present.

(2) If the fluid is encysted, it will not shift.
(3) Huge (massive) ascites does not shift. Do fluid thrill(wave sign) and not shifing dullness.
(4) Examine for enlarged abdominal organs by <u>dipping or ballottement</u> if the ascites is moderate or large.
(5) The distension may be due to solid masses such as a huge spleen, a huge liver (polycystic for example), huge paraortic lymph nodes, a pseudopancreatic cyst, huge polycystic kidneys, or a pelvic mass.

These should be assessed and a sound differential diagnosis should be given.

1. <u>A patient with generalised lymphoadenopathy</u>

Such patients may have history of anorexia, weight loss, fatigability, or fever.

2. Observe the general look of the patient (ill, cachexic, etc.)
3. Examine for: pallor, jaundice, clubbing
4. Examine the neck for lymphoadenopathy (all areas)
5. Examine for other enlarged lymph nodes (axillae, epitrochlear, inguinal, para-aortic).
6. Describe the enlarged lymph nodes(size, discrete or matted, tenderness, presence of sinuses, firm of rubbery).

If there is enlarged group of lymph nodes, examine the spleen, and para-aortic lymphoadenopathy.

Report your findings and give a rational differential diagnosis. That includes:

(1) Tuberculous lymphadenopathy
(2) Leishmaniasis, toxoplasmosis
(3) HIV infection
(4) Lymphocytic leukaemias (middle + old)
(5) Acute leukaemias (young)
(6) Lymphomas
(7) Sarcoidosis
(8) Connective tissue disease

Investigations depend on the cause, but start with complete blood count and ESR.

PPD (Mantorx) is strongly positive in TB lymphadenitis and negative or weak positive in malignancies and HIV

Altrasonography or CT of the abdomen

The ultimate diagnostic test is lymph node biopsy. Macroscopically.

Look for caseating granuloma that indicates TB.

Abdominal Masses (Not Ascites)

You may be asked to examine the abdomen of a patient with abdominal masses.

1- Approach the patient as usual.
2- Perform general examination noting the face of patient, pallor, jaundice, plethora (alcohol – polycytaemia, major venous obstruction).
3- Hands and nails to clubbing and other signs of cirrhosis.
4- Inspect the abdomen for visible swellings:
   Right upper quadrant – liver

Left upper quadrant – spleen

Epigastrium: stomach, pancreas left lobe of liver

Loins: enlarged kidneys (polycystic)

Central: para-aortic lymph adenopathy and retroperitoneal tumours.

Hypogastrium: distended bladder, ovarian, and uterine masses in women.

Right iliac fossa: tuberculosis, Crohn's, caecal masses.

Left iliac fossa: colonic masses

Proceed to examine for liver and spleen as explained earlier.

Difficult Situations

One of the difficult situations is to palpate for kidneys. An enlarged kidney due to polycystic disease can be a challenging situation.

1- A large left polycystic kidney can be easily confused with splenomegaly if examination was not properly performed.

2- Observe the fullness, whether it is confined to the loin or not.

3- Perform bimanual palpation, if appreciated, this is suggestive of kidney masses.

4- Try to put your fingers between the mass, and the left hypochondriumif possible it is the kidney.

5- Examine for a splenic notch.

6- Most importantly, perform careful percussion. A kidney mass is usually <u>resonant</u> unlike the spleen. Perform similar examination for the right kidney to differentiate it from an enlarged liver. Enlarged kidneys can be due to hydronephrosis (symmetrical), or renal malignancy (usually unilateral).

- If you are sure that you felt polycystic kidneys, request to examine for BP (hypertension), signs of chronic renal failure(A-V fistula).

Causes of Bilateral Renal Enlargement

Look for:

- Surgical scars
- AV fistula/haemodialysis
- Peritoneal dialysis catheter
- BP

Causes:

- Adult Polycystic Kidney Disease
- Bilateral hydronephrosis
- Amyloidosis
- Tuberous sclerosis with non-malignant neoplasms

Unilateral renal enlargement:

- Polycystic kidney(either with only one palpable or one removed)
- Mitotic lesion
- Unilateral hydronephrosis
- Hypertrophy of single functioning kidneyRenal cell carcinoma

Transplanted Kidney

- Scar in RIF
- Mass underneath scar – dull to percussion, possibly a bruit, non-tender (functioning)

Look for causes:

- Diabetes – insulin, marks on abdomen, needle marks on fingers, ophthalmoscopy
- Hypertension – ask for BP
- Adult polycystic kidney disease (bilateral renal masses or nephrectomy scars)

Look for previous replacement (dialysis):

- AV fistula – patent/ligated, recently needled
- Old Hickman/Portacath site
- Old Peritoneal dialysis scars

Look for complications:

- Uraemic? Requiring treatment (needled fistula)
- Ciclosporin signs (gum hypertrophy, hypertrichosis, skin lesions)
- Look for infections from immunosuppression
- Look for renal tenderness
- Renal bruit(renal artery stenosis)

Do not forget to auscultate the abdomen for bruits over the liver, and over the renal angles for the bruits of renal artery stenosis. Also, auscultate for venous hum over dilated veins.

Patients with Inflammatory Bowel Disease

Crohn's disease/Ulcerative colitis

- Young, thin patient with multiple abdominal scars. Could have a cutaneous fistula, ileostomy, or PEG feeding. He may have Cushingoid appearances, evidence of immunosuppressive therapy, or evidence of nutritional deficiency.
- Look for clubbing.
- Mouth ulcers, pyoderma gangrenosum, or erythema nodosum
- UC patient may have a lower laparotomy scar with either an ileostomy, or arepaired ileostomy site (with presumed IPAA). If jaundiced or signs of CLD–PSC.

Practical Gastrointestinal Cases

Examine this patient with abdominal distension.

The patient is a 51-year-old man lying comfortably in bed. He was of average height. His weight was difficult to judge as his abdomen was distended, but there appeared to be loss of weight as his extremities looked thin.

On examination of his head and face, his sclerae were icteric, making you immediately suspect that the patient had liver disease. There were multiple telangiectasia over his face (paper money sign) in keeping with advanced cirrhosis(alcohol related). There was bilateral parotid enlargement, which indicate that the aetiology of the condition was alcohol abuse.

On examination of the hands, there was palmar erythema. He had Dupuytren's contracture, which further suggest the aetiology of the condition was alcohol abuse.

He had multiple spider angiomata (naevi) over his upper chest a further indication of cirrhosis. There was no gynaecomastia and the distribution of hair growth over his chest was normal.

The abdomen was distended and the distension was greatest in the flanks, which indicate that the patient had ascites. There were visible veins on the anterior abdominal wall, and the direction of blood flow was from caudal to cranial(upwards in veins above the umbilicus, and downwards in the veins below the umbilicus). This suggested that the patient had portal hypertension.

By dipping (ballottement) technique of the abdomen (because of the ascites), the liver edge was detected 3 cm below the costal margin, but the spleen wasnot detected. Percussion demonstrated a horseshoe-shaped area of dullness and shifting dullness confirming presence of ascites. Liver span was 15 cm. The surface of the liver was nodular with a firm consistency. Auscultation did not reveal a bruitover the liver or venous hum. You suggested to examine the genitalia for signs of cirrhosis.

The likely diagnosis:

Advanced liver cirrhosis with portal hypertension.

Aetiology probably alcohol abuse. Possibly complicated with hepatocellular carcinoma.

Quick brief notes: Summary

Chronic liver disease

Incidental findings	Clubbing, tattoos
+ve signs	Icterus, Dupuytren's, palmar erythema, scratch marks, spider naevi, gynaecomastia, ascites. Shrinken liver and splenomegaly.
Differential diagnosis	Of the cause: hepatitis B/C; alcohol; autoimmune, metabolic, creptogenic?
Function	Compensated/decompensated: hepatic flap, encephalopathy, ascites
Tests	FBC(anaemia chronic disease) LFTs including GGT, alpha-fetoprotein, autoimmune and metabolic workup. Hep screen: viral serology, autoantibodies, iron+ferritin, AFP USS/CTabdomen

Causes of Cirrhosis:

Alcohol
Viral hepatitisAutoimmune: CAH, PBC
Metabolic: Haemochromatosis, Wilson's, α1AT

Notes:

- Defined as diffuse liver abnormality characterised by fibrosis and abnormal regenerating nodules
- Other causes: right heart failure, constrictive pericarditis, budd-chiari, Primary Sclerosing Cholangitis, galactosaemia, storage disorders
- Complications cirrhosis: portalhypertension, variceal haemorrhage, heaptic encephalopathy, ascites, hepatorenal syndrome
- Precipitants hepatic encephalopathy: Infection, diuretics, Upper GI haemorrhage
- Management variceal bleeding: ABC, blood transfusion, endoscopy, octreotide if no rsponse then consider TIPPS

Hepatomegaly/Splenomegaly

Splenomegaly

Incidental findings	Lymphadenopathy
+ve signs	Spleen enlarged at . . . cm, liver span > 12 cm
Differential diagnosis	Splenomegaly, Polycystic kidney. Causes below
Function	
Tests	FBC – for pancytopaenia Blood film – for ↓RBC/platelets/neutrophils USS/CT abdomen ±Bone marrow biopsy

Hepatomegaly

Incidental findings	CCF, cachexia
+ve signs	Liver is palpable, span >12 cm, ?Tender/smooth/ firm ?nodular. Listen for bruit
Differential diagnosis	Hepatomegaly, kidney; Causes below
Function	Stigmata chronic liver disease/decompensation
Tests	CBC, Liver function tests, hepatitis screening, autoantibodies, bone marrow, US/ CT abdomen...etc

Massive spleen	Moderate spleen	Liver	Liver + spleen
CML, hairy cell L	Myeloproliferative	Cirrhosis	Myeloproliferative
Myelofibrosis	Lymphoproliferative	Metastasis	Lymphoproliferative
(Malaria)	Cirrhosis	CCF	Cirrhosis
(Kala-azar)			TB

Notes:

- Causes of small spleen, same as moderate, with addition of infections (e.g., SBE, glandular fever).
- If hepatosplenomegaly and lymphadenopathy think of CLL/ lymphoma and infectious mononucleosis.
- Other causes hepatosplenomegaly: hepatitis, brucellosis, Weil's, Toxoplamosis, CMV, Pernicious anaemia, storage disorders, amyloidosis.
- Other causes splenomegaly: brucellois, sarcoid, ITP, Felty's, amyloid, Gaucher's (massive).
- $\Delta\Delta$ of hepatosplenomegaly is polycystic kidney disease.
- Myeloproliferative (CML, myelofibrosis, PCRV, essential thrombocythaemia).

- Lymphoproliferative (CLL, lymphoma, waldenstroms).
- Normal upper border liver is fifth IC midclavicular line.

Polycystic Kidneys

Incidental findings	AV fistula / PD catheter
+ve signs	Signs of uraemia, bilateral masses in flanks, ballotable, can get above
Differential diagnosis	PCKD, hepatosplenomegaly
Function	
Tests	FBC – polycythaemia; biochemistry, renal function Dipstick – microscopic haematuria ECG – LVH USS kidneys CT / MR angio to screen for intracerebral aneurysms

Reminder:

- Look for III nerve palsy of previous subarachnoid/aneurysm
- Ask to check the blood pressure

Notes:

- Can present hypertension, loin pain, haematuria, subarachnoid, infection
- 77 per cent patients with ADPKD reach end-stage renal faiure or die by age of seventy.
- Other features: Berry aneurysms, mitral valve prolapse, hepatic cysts (more marked in infantile, AR form)

- USS criteria for diagnosis in individuals at risk: Age <30 at least 2 cysts; 30-59 2 cysts in each kidney; >60 4 cysts in each kidney.

Renal Transplant

Incidental findings	Finger prick /lipo/atrophy/hypertrophy suggesting DM
+ve signs	This patient has ESRF, as evidenced scar in RIF with transplanted kidney palpable. ±Bilateral masses in flanks suggesting PCKD as aetiology. Renal replacement: Radiocephalic/brachiocephalic fistula or PD catheter/scar. Comment on whether fistula is functional and whether there is evidence of needling.
Differential diagnosis	If no specific cause found, 'the most common aetiologies inare: acute: paracetamol overdose', hepatitis, congenital.
Function	Rejection: 'The graft is non-tender, there are no scratch marks suggestive of uraemia' Fluid balance: JVP/basal creps/able to lie flat Immunosupression: 'There are purpora suggestive of steroid usage, multiple warts consistent with azathioprine and hypertrichosis and gum hypertrophy consistent with cyclosporine'.
Tests	

Reminder:

- Look for III nerve palsy of previous subarachnoid/aneurysm (polycysteic kidney)
- Ask to check the blood pressure

Causes of end-stage renal failure:

Diabetes mellitus
Hypertension
Glomerulonephritis
PCKD

Notes:

- Haemolytic uraemic syndrome and focal glomerulosclerosis: recurrence within graft common
- Graft survival post-renal transplant: HLA identical living 88% 5y, 70% 10y; cadaver / non-HLA identical donor ~70% 5y, 50% 10y
- Complications of transplant: infection (CMV, PCP), hypertension (ciclosporin), malignancy (lymphomas, skin cancer)
- Combined kidney-pancreas transplant prolongs survival in patients with diabetes and ESRF
- Acute rejection: Lymphocytic intersitial infiltrate, biopsy kidney, treated high-dose methyl pred, antilymphocyte globuline, OKT3
- Chronic rejection interstitial fibrosis, atrophy of tubules
- Chronic kidney disease stages 1-5 defined by GFR (15/30/60/90)

3. Examination of the Respiratory System

This can involve longand short cases.

Some examples:

1. Interstitial lung disease (most frequent by a long way!)
2. Bronchiectasis
3. Cystic fibrosis
4. Pneumonectomy/lobectomy

5. Pleural effusion
6. Chronic obstructive airways disease
7. SVC obstruction
8. Cor pulmonale
9. Apical fibrosis
10. Pleural effusion

Pleural Effusion:

1- This is a common problem, especially in undergraduate examinations.
2- Usually, the effusion is moderate or less (long effusion are usually aspirated).
3- Note the age of the patient (young: infection, pneumonia, TB, connective tissue disease; and the old: malignancies).
4- Proceed as follows:

Observe the position of the patient and his surroundings:

• Propped up, sitting up, tripod position (CPOD).

The scenario:

This patient, known to have COPD for ten years, presents to emergency department with a fever of 102.5°, cough, severe dyspnea. He states he is coughing up more secretions than usual, and that it is yellower and thicker. Vital signs showed a respiratory rate of 28/m, a heart rate of 122/m, blood pressure of 158/92, and an oxygen saturation of 89 per cent on room air. He is moderately obese. Upon assessment, he is sitting on the examination table bent forward, audibly wheezing, and using accessory chest muscles to breathe. The doctor auscultates his lungs and finds diminished bases and expiratory wheezes throughout all fields.

Patients with emphysematous COPD are typically thin but barrel-chested. They tend to breathe through pursed lips, and they sit leaning forward in a 'tripod position'. This posture widens the chest as much as possible by supporting the upper body on the elbows or the extended arms.

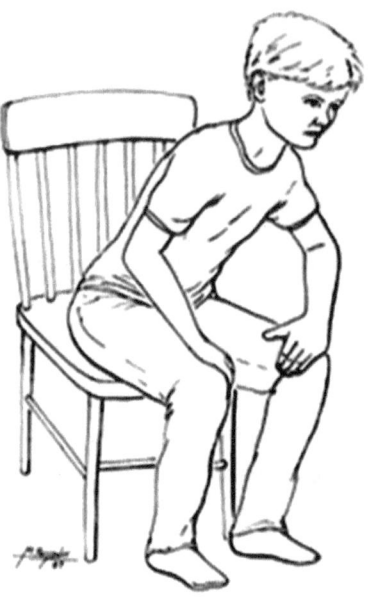

Fig 13. The tripod position. Patient leans forward, resting on elbows or hands in an effort to expand the chest and ease breathing. (Ref: Jason M. Alexander, MFA. © 2007, Wild Iris Medical Education.)

Further assessment:

Note any connections to machines as ventilator, nebulizer, O_2 masks, or nasal prons, chest tube, inhalers, etc.

- Note any sputum bottles by the colour of the sputum.
- Observe the breathing rate and character and type of breathing:
- Acidotic breathing : metabolic acidosis (hyperventilatoin)
- Deep and sighing: Kussmaul's (late phase of acidosis)
- Cheyne-Stokes: respiratory failure (and failure of other major organs)

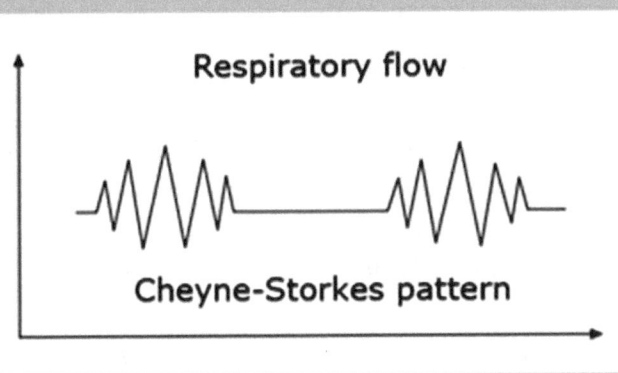

Fig 14: Cheyne-Stokes breathing pattern

- <u>Biot or irregular</u>breathing: it is described as the fibrillantion of respiration. It is seen in meningitis and severe depression of the respiratory centre.
- <u>Stertorous breathing:</u> harsh, noisy breathing usually heard in a comatose patient.
- <u>Hysterical:</u> hyperventilation (with tetany) or deep breathing with sudden holding of breath every five to six breaths.

Face and eyes:

o Horner's syndrome (lung CA in apex):

  - Ptosis
  - Miosis: partially constricted, but reacts normally to light
  - Anhydrosis: Place back of finger over each eyebrow to compare sweating

o Chemosis [tear that doesn't drop] (CO_2 retention)
o Eye fundus: papilloedema
o Conjunctiva: pale (anemia)

Mouth:

- Lips blue (peripheral cyanosis)
- Pursed lips breathing (emphysema, but not chronic bronchitis)
- Teeth: nicotine stains
- Teeth: broken, infected (predisposition to pneumonia or lung abscess)
- Tonsils: tonsils inflamed (upper RTI)
- Pharynx: reddened (upper RTI)
- Tongue: leukoplakia (smoking, spirits, sepsis, syphilis, sore teeth)
- Under tongue (central cyanosis)
- Voice: hoarseness (recurrent laryngeal nerve: lung cancer)
- Voice: stridor (upper airway obstruction)

Neck examination:

- Lymph nodes(tuberculosis, ca lung, lymphoma, connective tissue disease)
- Thyroid: Grave's disease, retorsternal goiter and tracgeal compression, & thyroid cancer)

Hands:

- Peripheral cyanosis
- CO_2 flapping tremor (CO_2 retention):
- Patient does a policeman 'stop' position with both hands
- Unlike liver flap, both hands go down at once
- Hypertrophic Pulmonary Osteoarthoropathy (lung CA)
- Erythema (CO_2)
- Tremor (asthma inhaler)
- Veins (CO_2)
- Muscle wasting of hands: inspect, then ask patient to adduct/ abduct againstresistance (brachial plexus palsy 2° to lung CA)
- Pallor of palmar creases (anemia 2° to blood loss)

- Pulse rate (asthma has tachycardia), rhythm, character, pulsus paradoxus (severe asthma). Large pounding pulse suggests CO_2 retension

Examination for Pleural Effusion

1- Go systematically: First, inspect for general appearance, surroundings, vital signs, chest appearance and movement, and breathing pattern. Then palpatefor trachea, tactile fremitus, apex beat (displaced by effusion), pleural rub (can be palpable), and chest expansion.
2- Percuss for dullness, especially stony (anterior and posterior). Do not confuse this with liver dullness on the right side(percuss during full inspiration), and cardiac dullness on the left.
3- You might find, above the area at dullness, an area of hyperresonance called skodiac resonance. This is a high-pitche dsoundelicitedbypercussingjustabovethesiteofapleuriticeffusion. This occurs if the effusion is moderate.

Auscultate carefully for:

1- Vocal fremetus using the 99-method.
2- Absent breathing sound over the area of stony dullness.
3- Absence of adventitious sound (crepitations) over same area.
4- Aegophony (nasal sound) can be heard over the upper limit o pleural effusion.
5- A severe form of bronchophony defined as a nasal or bleating quality of transmitted vocal sounds. Itis performed by asking the patient to say the letter *E*, while listening with the stethoscope to each lung field(when egophony is present, the sound is transmitted as *A*).
6- When you diagnose pleural effusion, mention the amount and the likely cause with other possible differentials.

Types of Pleural Effusion

Those are based on the history and presentation, as well as the colour and fluid analysis.

- Straw-coloured (serous):

  Indicated inflammation (bacterial or mycobacterial), connective tissue disease, heart failure, and nephrotic syndrome.

- Purulent (empyema)
- Haemorrhagic: malignancy, tuberculosis, pulmonary infarction, and blood diseases.
- Chylous: (a) due to injury or obstruction of the thoracic duct (by malignant lymph nodes, tumour of the lung or (TB); (b) filariasis (Bancroftion)

1- Be prepared to discuss the aetiology of the pleural effusion according to:

  a. History and examination: short history usually suggests parapneumonic. Long history, smoking, recurrence with weight loss, presence of lymph nodes, or evidence of secondaries indicates malignancy.

2- Clubbing is associated with empyema, longstand TB, lung abscesses, and malignancy(painful due to hypertrophic pulmonary arthoropathy).
3- Mention how analysis of the pleural fluid will help: protein content (exudate or transudate).

Fluid is exudate if one of the following Light's criteria is present:

1- Effusion protein/serum protein ratio greater than 0.5
2- Effusion lactate dehydrogenase (LDH)/serum LDH ratio greater than 0.6

3- Effusion LDH level greater than two-thirds the upper limit of the laboratory's reference range of serum LDH
4- Empyema if pH >0.6
5- Excess lymphocytes in the fluid is suggestive of:

TB
Lymphoma and malignancies
Viral pneumonias

6- Excess polymorphs indicate bacterial infection.
7- Excess eosinophils suggest filariasis and makes diagnosis of malignancy unlikely.

What tests to be done for pleural effusion:

o   Adenosine deaminase:

- To confirm TB effusion if> 40 U/L. Also, this level suggests empyema in 60 per centand complicated paraneumonic effusion in 30 per cent.
- Glucose < 60 mg/dl (3.3 mmol/L suggests complicated paraneumonic, TB, and empyemaeffusion. Also occurs in 20 per centof malignanat effusion.
- Lactate dehydrogenase: If > than 2/3 rds of upper limit of normal, or ratio of fluid to serum > 0.6is suggestive of exudatative effusion, especially complicated paraneumonic one.
- RBC > 100,000 (haemorrhagic effusion) indicates malignanacy, pulmonary embolism, or paraneumonic effusion.
- WBC > 10,000 is not a good indicater of purulent fluid as pus contains most of dead WBCs and debris. Neutrophils > 50 per cent suggest parapneumonic effusion. In 7 per cent of TB cases.
- Lymphocytes > 90 per cent occurs in TB and lymphoma.

Difficulties:

A small pleural effusion can be confused with pleural thickening. In this case, the dullness is not typically stony and confined to a small area.

Patients with Consolidation and/or Fibrosis

If you are asked to examine the chest of a patient, and after general examination (mentioned previously), you inspected the chest. Then ifone apex is depressed or flat, this is suggestive of long-standing fibrosis (or collapse). In this situation, proceed and observe:

- The movement with deep breathing (expected to move less)
- Confirm this by palpation (the chest expansion should be less than 2cm to be significant).
- Palpate for the trachea: in fibrosis and collapse, it will be pulled towards the side of the pathology.
- Percussion note will be impaired and breathing sounds will be diminished with evidence of bronchial breathing.
- If you hear crepitations and bronchial breathing, these suggest presence of fibrosis and consolidation (old TB with recent reactivation).
- Fibrosis without consolidation causes bronchial breathing without crepitations.
- If there is no depression in the upper chest, but one side of the chest moves less, this suggests consolidation and proceeds to confirm the movement and position of the trachea by palpation. Also, elicit tactile and vocal resonance (increased in consolidation).
- Listen also for diminished breathing sounds, crepitations, and bronchial breathing. Determine the quality of bronchial breathing by looking for:

1. High-pitched (tubular) indicates consolidation.

2. Low-pitchedthat is also called canvernous(amphoteric if like blowing over the neck of the jar) is suggested of cavitation.

3. Elicit increased vocal resonance and perform whispered pectoriloquy by asking the patient to whisper ninety-nine or one-one-one or in arabic 44 (arbawarbaeen). It is present when you hear the whiper loud enough, as if it is whispered directly on your ear.

- Differential diagnosis of collapse fibrosis alone or with crepitation:

The history is very important as the duration and the illness, age of the patient, and presence of fever and weight loss are important to decide what pathology is present.

- In clinical stations, history is not allowed; and therefore, make a differential according to the findings.
- Collapse in a youngpatient may be caused by:

  o Foreign body inhalation
  o Obstruction of bronchi by enlarged lymph nodes to TB, or lymphoma.
  o Thick mucus plugs. While in older patient, it is likely to be dueto obstruction by tumour, (ca bronchus) or malignant lymph nodes.

- Fibrosis in one area (say one apex) is usually due to long-standing inadequately treated TB, bilateral fibrosis of upper lobes is due to cystic fibrosis (also ankylosing spondylitis), but fibrosis in lower lobes is often due to bronchiectesis and interstitial lung disease.

Physical Signs of Collapse Compared to Fibrosis

Collapse	Fibrosis
Flatteningnot resent	Present ++
Trachea shifted to same side (Upper lobe)	Pulled to same side (upper lobe)
Mediastinum: deviated same side in lower lobe collapse	Deviated to same side if gross
Percussion impaired	Less impaired
Dimished vocal resonance	Normal or decreased
Breathing sounds decreased	Decreased
Bronchial breathing +	- (+)
Crepitations+	- (+)

Causes of fibrosis:

(1) Apical: TB (rarely lower lobes), ankylosing spondylitis
(2) Lower lobes: bronchiectesis (can involve upper lobes)
(3) Pneumoconiosis (bilateral)
(4) Sarcoidosis
(5) Interstitial pulmonary fibrosis:

- Idiopathic
- Connective tissue disease
- Drugs: cytotoxic, amiodarone

Example:

Examine this man who has progressive shortness of breath.

- The basics: wash hands, greet, introduce self, confirm patient name, and tell the patient that you will examine his respiratory system if that is allowed. Tell him that if he has any painful area in her chest.

Action:

General examination: look: from end of bed and the right side.

The patient was an elderly male resting comfortably in bed reclined at approximately forty-five. He was of average height and weight. Comment on surroundings: oxygen mask, no sputum pot, or inhalers.

> *Better to start with head and face: several clues in face will direct you to have the right diagnosis.

On examination of his head: Look for any use of accessory muscles such as the sternocleidomastoid muscle. Also palpate for the left supraclavicular node (Virchow's Node). Eye examination was unremarkable. No facial rash or parotid enlargement. Look for puffing through pursed lips(airway obstruction, emphysema). You also noted central cyanosis.

At the wrist, you should take the patient's pulse. A bounding pulse with warm palms may indicate carbon dioxide retention. On examination of his hands, look for small muscle wasting(pancoast tumour). Also look for nicotine staining (COPD or lung cancer).

You noted that the patient had clubbing (increased nail bed fluctuation, loss of the nail be angle, increased curvature of the long axis of the nail, stage three clubbing. At this point, recall the causes of clubbing with special reference to the causes in relation the respiratory system.

Findings: There were no other abnormalities detectable on examination of the hands, especially no fine tremor(carbon dioxide retention or asthma and Beta-2 agonists use) and no nicotine staining.

There were no abnormalities detected on examination of the neck. The trachea was in the midline. On examination of the chest, you found that the chest was normal in size and shape with normal respiration with equal movements of the two sides of the chest.

The respiratory rate was twenty-four per minute; the apex beat was in the fifth left intercostal space just medial to the midclavicular line(proper technique). Vocal fremitus was normal and equal on the two sides; respiratory movements by palpation were equal on the two sides over upper, middle, and lower chest anteriorly and posteriorly.

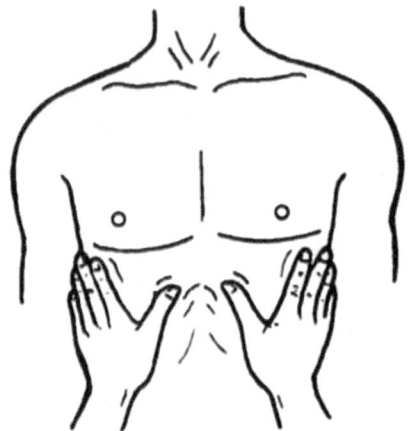

Fig 15: Performing chest expansion

Percussion note was resonant and equal on the two sides. Breathing sounds were vesicular. There were fine late inspiratory crepitations at both lung bases. This made the one consider the causes of fine crepitations at the lungs bases. You thought of fibrosing alveolitis. In association with clubbing, this was the most likely diagnosis. Vocal resonance was normal and equal on the two sides.

Diagnosis:

This patient has fibrosing alveolitis, not in respiratory failure clinically. Next, we considered the questions that could be asked.

Discussion:

First, causes of fibrosing alveolitis. Next, whatinvestigations that one may arrange on this patient (high resolution CT, pulmonary functiontests). Finally, consider management plans.

Physical Signs of Cavitation

(1) Diminished movement
(2) Cracked pot note on percussion
(3) Cavernous or amphoteric bronchial breathing
(4) Crepitations
(5) Whispering pectoriloquy

These signs depend on the site of the cavity, its proximity to the surface, and the content of fluid inside it.

Other respiratory problems:

(1) <u>The patient with chronic obstructive pulmonary disease.</u>

   In general, such pateints present with complications such as:

   a.   Cor pulmonate (old description: blue bloaters)
   b.   Emphysema (old description: pink puffers)

(2) Observe the face of the patient and his position in bed (tripod position), and note oxygen devices or nebulizers.
(3) Look for cyanosis.
(4) Observe the accessory muscles of respiration (usually in use). The abdominal respiratory effort is usually decreased.
(5) Involuntary pursing of the lips is often seen during expiration to lessen the airway obstruction.
(6) Observe the respiratory rate and the prolonged expiration. Also, the movement of the chest.
(7) Examine the hands for:

   • Nicotine staining
   • Flapping tremor, warm handsindicating carbon dioxide retention
   • Clubbing: if present, suspect an additional pathology such as bronchiectesis or lung cancer(painful).

(8) Examine the JVP: if raised, look for other signs of right side heart failure.

(9) Observe the shape of the chest: barrel chest, which occurs in emphysema.

(10) Examine for tracheal tug(also called campbell's sign): downward movement of the thyroid cartilage in COPD.

(11) Attempt to measure the cricosternal distance. Normally, it is around three fingers. It is reduced in COPD.

(12) Oliver's sign is downward movement of cricoids cartilage with each heart beat present in aortic arch aneurysm.

(13) Summary of physical signs of COPD:

- A reduction in the length of the trachea palpable above the sternal notch
- Tracheal descent with inspiration: Tracheal tug
- Excessive use of the scalene and sternomastoid muscles
- Excavation of the suprasternal and supraclavicular fossa during inspiration
- Jugular venous filling during expiration
- Loss of buckethandle movement of the upper ribs
- Paradoxical movement of the lower ribs
- Prolonged forced expiratory time.

Other abnormal shapes:

(14) Pigeon chest Pectus carinatum (with kyphoscoliosis). Occurs in childhood rickets or chronic respiratory infections.

(15) Funnel chest (pectus excavatum): depression of the lower sternum and is either congenital or due to rickets.

(16) Examine properly for chest movement by palpation.

- Note that the normal chest expansion at the maximum circumference is at least two inches (5cm).

(17) Conditions causing bilateral decreased expansion.

1. Emphysema
2. Ankylosing spondylitis

(18) Conditions causing unilateral diminished chest expansion:

- Pleural effusion
- Pneumothorax
- Fibrosis
- Consolidation
- Atelectasis

Other short cases in the respiratory system stations:

(19) Some patients may have mixed pathologies or minimal abnormalities. Thus, meticulous and careful examination should be performed.

Examples:

Patient with hydropneumothorax:

- Trachea may not be pushed to opposite side.
- Percussion is not expected to be hyperresonant. It may be dull in the lower areas and resonant at the top.
- Breathing sounds will be diminished or absent (also tactile and vocal fremitus).
- You may hear succession splash (hippocratic succession) on shaking the patient and listening to the chest.

Mixed Respiratory Conditions

(20) A patient may have consolidation plus collapse, or consolidation plus fibrosis.

This often occurs in the apices and one should be careful during examination and look for:

- Presence of apical depression favours fibrosis if long-standing.
- If the trachea is pulled towards the affected side, this suggests also significant collapse or fibrosis.
- Other signs are shared by collapse and fibrosis (impaired percussion, tactile and vocal resonance, and breathing sounds). But fibrosis can cause increasedfocal resonance, bronchial breathing, and crepitations.
- Presence of bronchial breathing with or without crepitations is in favour of consolidation, though fibrosis (and rarely collapse) can give such signs.

(21) <u>Patient with pleural thickening</u>

It is relatively difficult to detect signs of thickened pleura, as those are usually slight and confined to a small area (bases).

- All modalities of examination are impaired, but there is no shifting of the trachea.

Mixed Respiratory Problems

Example:

76-year-old gentleman with instruction, 'This man has had thoracic surgery and is now increasingly short of breath.' He had a big midline sternotomy scar, along with a scar in axilla and on the right thorax posteriorly, which I presented as consistent with operative drains. He had marked fibrosis to midzone on left side with reduced percussion basally with normal sounds on right.

Likelydiagnosis: end-stage fibrosis with a single lung transplant. Discussed concerns recurrence of bronchiolitis obliterans in lung transplants, the need to monitor spirometry and discussed the complications of immunosuppressants.

Lobectomy or Pneumonectomy

Pay attention to breathing sounds entry during auscultation.

o   If air entry is reduced, then think of lobectomy and pneumonectomy.

o   If air entry is equal, then think of decortication and bullectomy.

Lobectomy vs Pneumonectomy

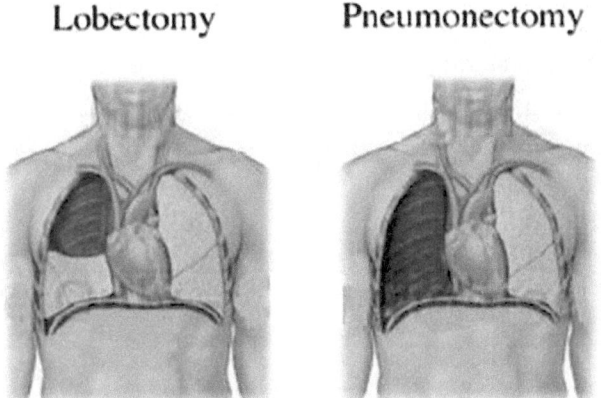

Fig 16: **Lobectomy vs Pneumonectomy**

The similarities:

1.   Both have thoracotomy scars.
2.   Both have reduced chest expansion and reduced AE.

The differences:

1.   The signs of lobectomy are confined to lobe, which is removed. The signs are similar to *pleural effusion* except it is, not stony dull.
2.   The signs of pneumonectomy are extensive(i.e., involve the whole lung. The side involved would be flattened. It is similar to *whole lung collapse*).

3. Normally, the tracheal iscentral in lobectomy except for upper lobe. The tracheal is almost alwaysshifted in pneumonectomy.

Indications for lobectomy:

- Lung cancer: 25 per cent of non-small cell lung cancer will be suitable for attempted surgical resection.
- Solitary pulmonary nodule of uncertain cause
- Localised bronchiectesis
- TB: was previously treated surgically

When you find a lateral thoracotomy scar in respiratory station, always think of three possibilities, namely lobectomy, pneumonectomy and previouslung transplantation surgery.

In case of lung transplant(or any organ transplant), always pay attention to look for side effects of cyclosporine, as well such asgum hypertrophy and excessive hair growth.

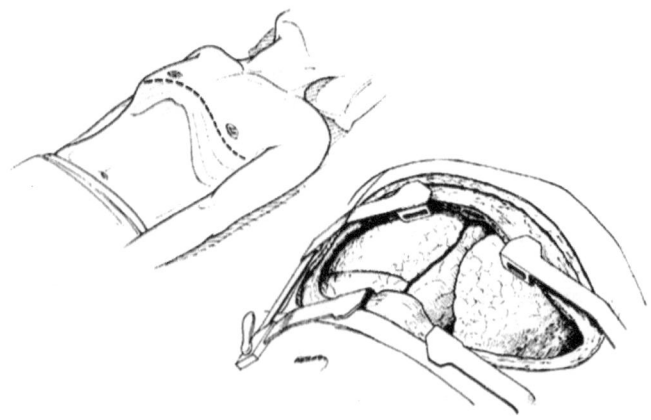

Fig: 17. Transverse thoracosternotomy, or 'clamshell' incision for bilateral lung transplant

- <u>Paralysis of the diaphragm: signs</u>

  By inspection: the lower costal margins move downwards from midline (opposite to the normal).

- If you place one hard over the lower chest and the other over the epigastrium, there is paradoxical retraction of the epigastrium during inspiration if there is diaphragmatic paralysis.
- Impaired percussion posteriorly with diminished or absent breath sounds and tactile and vocal resonance.
- <u>Tidal percussion:</u>

Determine the lower limit of lung resonance during inspiration and expiration. Normally, there is a difference of 5cm.

Bronchiectasis

- Lower lobes affected more than the upper
- Patient may be tachypniac, cyanosed, or febrile.
- Finger clubbing is usually present.
- Percussion note and tactile vocal resonance are normal or diminished. Bronchial breathing is not usually heard unless there is surrounding consolidation.

Numerous coarse crepitations are the main findings on auscultation. Biphasic crackles occur in both inspiration and expiration. They are a combination of coarse and fine crackles. Biphasic crackles are a feature of bronchiectasis and are related to a combination of secretions and increased compliance of the walls in larger airways.

- Crackles and rhonchi are often observed in association with active infections and acute exacerbations.
- Apical bronchiectasis is usually due to TB(dry bronchiectasis with little sputum)

(1) Do not confuse interstitial lung fibrosis with bronchiectasis as in both clubbing is found. However, both conditions are associated with cyanosis, although it is more seen with interstitial fibrosis.
(2) <u>Interstitial fibrosis</u> the crepitations are usually late inspiratory, while in bronchiectesis, with copious purulent sputum, are coarse and therefore do not occur in fibrosing alveolitis.

(3) <u>Remember:</u>

Bronchiectasis can complicate cystic fibrosis or congenital dextrocardia (kartagener's syndrome). It is also associated wit chronic sinusitis. Septic embolic can lodge in the brain and cause brain abscesses.

Unilateral bronchiectasisshould alert you to the possibility of foreign body or obstruction of the bronchial tree.

Pneumoconiosis

(1) It is rare to see patients with pneumoconiosis in daily practise or examinations.

The <u>symptoms</u> of both silicosis and asbestosis are those of dyspnea, cough, with or without purulent sputum.

(2) The physical signs are those of bronchitis and emphysema or superadded infection.

(3) If complicated by cor pulmonale, one will find signs of such disease: raised JVP, lower limb oedema, and evidence of right ventricular hypertrophy.

(4) <u>Remember:</u> asbestosis can be complicated by lung or pleura malignancy (methosilioma).

Other Respiratory Examination Cases

Examine this patient who has cough and feels unwell.

The patient was a 72-year-old man. He was seated up in bed and had an oxygen mask on. He was of average height, but he looked thin. His skin was pigmented.

(Suggestion: patient had a problem involving the respiratory system because of the breathlessness, and it was probably a chronic illness because of the emaciation).

Also, think of the causes of skin pigmentation like Addison's disease in association with tuberculosis, or he may have non-metastatic manifestations of bronchial carcinoma.

On examination of the head, we noticed that the head was verythin andreinforcing the impression of emaciation and chronic illness. No lymphoadenopathy was found.

The alae nasi were flaring, indicating that the patient was in respiratory distress. He has conjuctival pallor, but no jaundice. Tongue is pale and smooth(iron deficiency anaemia).

There was no specific abnormality noted on examination of the hands, particularly no clubbing, nicotine staining, or painful swollen risks.

On examination of the neck, we noted that the neck was quite. The trachea was deviated to the left hand side.

Chest examination:

The chest wall was thin and almost skeletal. There was gynaecomastia. The respiratory rate was twenty-eight per minute. Respiratory movements were decreased on the right chest. Tracheal deviation to the left and reduced movements on the right would suggest either pleural effusion or pneumothorax on the right side. However, the illness was a chronic process, and this suggests the diagnosis was pleural effusion.

Taking into account the emaciation, pigmentation, and gynaecomastia the cause of the effusion was most likely a bronchial cancer.

Vocal fremitus was decreased on the rightside in keeping with the suspicion of pleural effusion on that side. Percussion note was stony dull on that side in keeping with pleural effusion. Breathing sounds were decreased on the rightside, and the vocal resonance was reduced, and there was aegophony at the upper level of the effusion

Diagnosis: Right-sided pleural effusion underlying bronchial cancer.

Concise Summary of Some Respiratory Problems

Bronchiectesis

Incidental findings	Clubbing, sputum pot (three layers)
Important symptoms and signs	Productive cough of large volume, smelly sputum, basal crepitations
Differential diagnosis	Bronchiectesis, ca lung, fibrosing alveolitis, abscess
Function	Respiratory compromise
Tests	CXR, Arterial blood gases HRCT

Causes:

Post-infective: pneumonia/TB
Bronchial obstruction: TB, carcinoma, lymph nodes
Hypogammaglobulinaemia
Cystic Fibrosis
Kartaneger's

Further discussion:

Look for splenomegaly (amyloidosis)

Notes:

- Other causes: bronchial obstruction, Marfan's, yellow nail
- Complications: pneumonia, haemoptysis, brain abcess, amyloidosis
- Pathogens: S.aureus, H.influenzae, P.aeruginosa
- Kartaneger's (ciliary dysmotility, dextrocardia, situs invertus, dysplasia sinuses, infertility)

COPD

Incidental findings	Warm palms, bounding pulse indicate CO2 retention. Clubbing suggests lung cancer. Nicotine staining: smoking.
+ve signs	Tracheal tug, hyperinflation, accessory muscles, hyperresonant, quiet breath sounds (prolonged expiration, wheezing)
Differential diagnosis	COPD
Function	CO2 retention, pulmonary hypertension
Tests	FBC – polycythaemia ABG – hypoxia, CO2 retention PFT – obstructive, reduced gas transfer CXR –signs of over-inflation, rule out infection/ mass lesion

Causesand types of lung injury:

- Smoking (mixed)
- Alpha 1 antitrypsin (panacinar)
- Coal dust (centrilobular)

Further findings: Look for splenomegaly (amyloidosis).

Notes:

- Emphysema – abnormal permanent enlargement of airway distal to terminal respiratory bronchiole.
- Chronic bronchitis – productive cough at least three months a year (two sucessive years)
- Indications long-term oxygen theraoy (LTOT): FEV1<1.5, P02<7.3, not smoking.
- Indications lung transplant (in general): <60, end-stage lung disease, life expectancy of twelve to eighteen months

Fibrosing Alveolitis

Incidental findings	RA, systemic sclerosis
+ve signs	*Clubbing*, dyspnoea, tachypnoea, cyanosis, fine end inspiratory bibasal creps
Differential diagnosis	Fibrosing alveolitis/bronchiectesis
Complications	Pul HT: ↑JVP, loud P2, RV heave
Tests	↑ESR, ANA, RF CXR – basal reticular shadowing ABG – hypoxaemia PFT – restrictive, desaturation after exercise HRCT ± lung biopsy

Causes

Upper	**Middle**	**Lower**
Extrinsic allergic alveolitis	Sarcoid	Idiopathic
Ankylosing Spondylitis	Beryllium	Connective tissue disease
Old TB		Pneumoconiosis
		Drugs

Further discussion: Look for splenomegaly (amyloidosis)

Notes:

- Connective tissue diseases(RA, systemic sclerosis, SLE, polymyositis/dermatomyositis, ankylosing spondylitis—upper lobe). In RA, fibrosis can also be caused by methotrexate.
- Drug causes(methotrexate, amiodarone, nitrofurantoin)
- Types of pneumonitis: Usual Interstitial Pneumonia (UIP), desquamative IP, Non-specific IP, lymphoid IP, giant cell IP
- Rx: trial of steroids for all, but taper if no response.

Pleural Effusion

Incidental findings	Clubbing, radiation burns, aspiration marks, lymphadenopathy
+ve signs	Stony dull percussion note, ↓or absent breath sounds, ↓vocal resonance
Differential diagnosis	Pleural effusion cause depends upon transudate or exudate
Function	Respiratory compromise
Tests	CXR, USOUND Pleural tap: protein, albumin, LDH, glucose, cyto, micro, AFB, adenosine pleural biopsy, CT chest

Causes

Transudate	**Exudate**
Cardiac failure	Neoplastic: Cancer bronchus/met/
Nephrotic syndrome	mesothelioma
Cirrhosis	Infectious: Pneumonia, TB
	Connective tissue: RA, SLE, PE

Notes:

- Protein > 30 = exudate, protein < 30 = transudate
- Lights criteria for exudate: (i) pleural: serum protein > 0.5; (ii) pleural: serum LDH > 0.6; (iii) pleural LDH > ⅔ upper limit for serum LDH
- Serum: effusion albumin gradient 1.2g/dl very specific for exudates
- Can have area of bronchial breathing above dullness. Due to. Consolidation: bronchial breathing, increased vocal resonance; Collapse: trachea to side, absent breath sounds.
- Meigs syndrome: ovarian tumour with right effusion (transudate)

4. Examination of the Central Nervous System

General Approach

Doctors and students may find difficulty in examining the CNS for a number of reasons.

(1) Like the examination of other systems, the CNS needs detailed history with emphasis on chronological order of events. Also needs good observation, meticulous and skillful examination, and logical interpretation of findings.

Reporting the diagnosis:

(2) In both long and short cases, give a full diagnosis that includes:

   a. The clinical diagnosis: e.g., this patient has a right-sided upper motor neuron paralysis involving the face, and both upper and lower limbs with G3 weakness;or this patient has lower motor neuron lesion of both lower and upper limbs.
   b. The anatomical location: e.g., this is most likely due to a left-sided cerebral hemisphere lesion, especially involving the internal capsule.

    c. The pathology: e.g., I think this patient has sustained cerebral haemorrhage, in view of being hypertensive, the sudden onset, the severe headache and vomiting, and the altered level of consciousness.

    d. The aetiology (cause of the suggested pathology): hypertension, or A-V malformation.

This cerebral haemorrhage is most likely caused by the poorly controlled hypertension (or ruptured A-V malformation).

This will make a <u>full</u> neurological diagnosis. After that, you can talk about the management and prognosis.

Other hints:

Be observant, look at the patient, and observe:

1. The level of consciousness: grade this according to Glasgow coma scale or Grady scale or minimal status examination (see appendix).
2. The position of the patient in bed and the position of limbs: upper limbs flexed and lower limbs extended in UMN (pyramidal) weakness (also lateral rotation of legs).
3. In some patients, the head is deviated towards the side of the lesion in UMN paralysis.
4. Examine the eyes for conguate gaze, which is the inability to move both eyes in a horizontal or vertical direction for fixed lateral or upward gaze. Stroke in the pons is a common cause of horizontal gaze paralysis ipsilateral to the lesion. Lesions in the contralateral cerebral hemisphere also cause horizontal gaze loss.

Vertical gaze loss usually occurs from midbrain lesions due to a lesion of the superior corpora quandrigemina. The pupils may be dilated with vertical nystagmus.

Spasm of lateral gaze or conjugate deviation also occurs at thestart of an epileptiform attack.

(3) <u>Paralysis of lateral gaze:</u> here, the patient cannot move his eyes to either right or left (paralyses of <u>lateral rectus </u>of one eye and <u>medical rectus </u>of the other).

It is due to a lesion of the upper pons (vascular or neoplastic).

Examples of Neurological Cases

The commonest long and short cases are:

- **Cranial Nerve Palsies**

Examples:

Olfaction (CN I)

Not usually tested unless a subfrontal brain tumour is suspected. Ask the patient to smell common odours like coffe or soap. Avoid using noxious (very strong) odours since they stimulate pain (CN V).

Testing for Vision (CN II)

a. **Visual acuity.** Usually tested using an eye chart. Test each eye seperately.
b. **Colour vision.** Also test eyes seperately using the red colour to test red desaturation, which can occur in optic neuritis.
c. **Visual fields.** This can be tested by fixing the head and asking the person to identify moving fingers in each quadrant of the field for each eye seperately.

More accurate testing can be done in the lab for patients who need regular follow-ups. If the patient is comatose, use the blink-to-threat technique where the examiner's fingers are rapidly moved towards the eyes of the patient from diferent quadrants to see any blinking.

d. **Visual Extinction.** This is suspected usually in cases of hemineglect. It is tested by asking the patient to count how many fingers seen when fingers are presented to both sides at the same time. If there is visual extinction, patient do not see the fingers on the affected side, but they can see the fingers if presented to the affected side alone. Visual extinction is seen in lesions of the contralateral parietal lesions.

Interpretation:

Lesions in front of the optic chiasm in the eye, oroptic nerve, cause visual deficits in **one eye**, while lesions behind the optic chiasm (optic tract, thalamus, white matter, visual cortex) cause **visual field** deficits that are similar for **both eyes**. For example, bitemporal hemianopia occur in pitutary tumours.

Pupillary Responses (CN II, III)

Look at the eyes and record the pupil's size and shape. Bilateral small pupils suggest morphine toxicity or bilateral pontine lesion. It occurs also in Horner's. Unilateral small pupil is causes by a unilateral pontine lesion.

A bilateral small irregular pupil that do not react to light, but eact sluggishly to accommodation, is called Argyll Robertson pupils, which is caused by neurosyphilis.

Bilateral dilated pupils occur in anticholinergic or tricyclic overdose. Bilateral dilated and fixed pupil is a sign of brain death. A unilateral dilated pupil occurs in third nerve palsy together with ptosis and deviation for the eyeball medially and downwards. Aulilaterally dilated pupil that react slugishly is called Adie's pupil.

Interpretation:

Impairment of the direct response in:

1. Ipsilateral optic nerve
2. Pretectal area
3. Ipsilateral parasympathetic in CN 111
4. Constricter muscle of iris

Impairment of consensual response:

o Contralateral optic nerve
o Pretectal area
o Ipsilateral parasympathetic in CN 111
o Constricter muscle of iris

When is accommodation response impaired:

• Ipsilateral optic nerve
• Ipsilateral parasympathetic travelling in CV111
• Constricter muscle of pupil
• And in bilateral lesions of the pathways from the optic tracts to the visual cortex.

Extraocular Movements (CN III, IV, VI)

Ask the peson to look at all directions as pointed by your finger without moving the head. Then ask if they see double vision. To test for convergence, ask the person to fix on an object as it moves slowly towards a point between the patient's eyes. Look also for nystagmus.

Testing for the oculocephalic response:

1. If there is no head or cervical spine injury, hold the eyes open and rotate the head from side to side or up and down. The eyes should move in the opposite direction of the head if the reflex is present(doll's eye).

2. The absence of doll's eyes suggests brainstem dysfunction in the comatose patient, but can be normal in the awake patient.

Interpretation:

- This test detects abnormalities in occulomotor, trochlear, and abducent from the brainstem to the orbit, or in the high order centres in the cerebral cortex.
- Spontaneous nystagmus is caused by toxic or metabolic conditions such as drug overdose or alcohol intoxication, or peripheral or central vestibular dysfunction.

Facial Sensation and Muscles of Mastication (CNV)

How to test:

1. Use a cotton whisp to test for facial sensations and use double simultaneous stimulation to test for tactile extinction.
2. The corneal reflex: This is mediated by the fifthand seventh. Use the cotton whisp to gentlytouch each corneaand observe for the asymmetry of blinkingcornea.
3. Testing for the motor function of fifth (trigeminal) is done by feeling the **masseter muscle when the jaws are clinched.**
4. Also, the jaw reflex is elicited by gently tapping on the jaw, while the mouth is half-open.

Interpretation:

- Impairment of the facial sensations in lesions of:

  o Trigeminal nerve
  o The trigeminal sensory nuclei in the brainstem
  o Or by lesions of the ascending sensory pathways to the thalamus and somatosensory cortex in the postcentral gyrus.
  o The corneal reflex is impared by lesions of the fifth, seventh, or their connections in the brainstem.

o Muscles of masstication are weak by lesions of the UMN pathways of the motor nucleous of the fifth, the lower motor neuron of the fifth in the pons, orin the neuromuscular junctions or the muscles themselves.

o The jaw reflex in brisk in bilateral UMN pathways connecting the trigeminal motor nucleous.

Facial (Seventh) CN

A 75-year-old male was admitted to the hospital because his daughter noticed that he woke up with a left facial droop and slurred speech. He has end-stage renal disease on hemodialysis (ESRD on HD), hypertension (HTN), diabetes type 2 (DM2), severe Aortic Stenosis (AS), 0.6 cm2, deemed inoperable due to ESRD.

Examine for seventh nerve to find if it is UMN (TIA/stroke) or LMN seventh palsy (Bell's).

The examination of CN 7 palsy can be remembered by the mnemonic **COWS**:

C lose your eyes
O pen (the examining physician tries to open the patient's eyes)
W rinkle your forehead
S mile

Don't ask 'show me your teeth' because a common reply is 'I don't have any teeth!'. This will test for the elevator angularis (NOT orbicularis oris).

• To test for orbicularis oris, ask the person to whistle, or give a straw and ask him/her to draw fluid.

Causes of seventh nerve lesions:

• UMN lesion is caused by contralateral cerebral hemisphere or descending central pathway. LMN lesion is caused by ipsilateral nucleus, nerve fibres, neuromuscular junction, or the muscle itself.

What is another additional important point on the physical examination of an ESRD patient?

Where is the HD access? What is the status of the access? In this patient, the access was a Left Upper Extremity (LUE) AV fistula with a good thrill and pulse. A thrill is the sensation that is felt over the anastomosis, typically described like a buzzing or vibration under the skin.

Hearing and Vestibular Sense (CN VIII)

Testing for Hearing

Whisper words or rub your fingers near each ear and ask the patient if these can be heard. To distinguish neural from conductive hearing problems, use the tuning fork. Rinne test is done by holding the vibrating toning fork out the ear to test for air conduction. Bone conduction is performed by placing the tuning on each mastoid process.

In Weber test, the tuning fork is placed in the midline skull. If there is sensorineural loss, the tone is louder in the normal side, while in conductive loss, the tone is louder on the affected side.

Air conduction is heard better by normal people. When bone conduction is better than air one, this indicates conductive hearing loss.

If air conduction is greater than bone one, this occurs in sensorineural loss as in normal people, but hearing is decreased in the affected ear.

Interpretation:

Significant unilateral hearing loss is caused by either peripheral neural or mechanical lesions.

This can be caused by diseases of the vestibular apparatus of the inner ear, the vestibular portion of eighth cranial nerve, the vestibular nuclei in the brainstem, the cerebellum, or pathways in the brainstem (such as the medial longitudinal fasciculus) that connect the vestibular and oculomotor systems.

Palate Elevation and Gag Reflex (CN IX, X)

To test for CN ninth and tenth, ask the person to say 'Aah'. The normalpalate will summetrically elevate. If there is a lesion or CN ninth, the affected side will not elevate (curtain movement).

Testing for CN tenth is done by brushing the posterior to elicite the gag reflex. Failure of the gag in patients with brainstem pathology, impaired consciousness, or impaired swallowing.

Muscles of Articulation (CN V, VII, IX, X, XII)

These nerves control articulation, which producethe normal speech. Ask the personif his speech is changed from baseline. Abnormal speech include:

- Heavy (cerebellar of extrapyramidal)
- Slurred(stroke)
- Nasal(cerebellar or brainstem)
- Intrrupted or stacatto (cerebellar, extrapyramidal)

Distiguish speech problems from abnormality of language production or comprehension, which cause **aphasia.**

Interpretation:

Abnormal articulation causing abnormal speech is caused by abnormaltities in muscles of articulation. The peripheral central portion of CN 5, 7, 9, 10, or 12.

Other disease that cause abnormal speech production include lesions of:

- Motor cerebral cortex
- Cerebellum
- Basal ganglia
- Descending pathways to the brainstem

Cranial Nerve Eleventh: Trapezious and Sternomastoid Muscles

Damage to this nerve causes the following symptoms:

- Pain in the shoulders
- Winging of the scapula
- Weakness and pain of the trapezius muscle

Examination:

- Observe the contour and bulk of each sternomastoid and trapezius, and look of winging of the scapula when the person pushes his hands forward against the wall.
- Ask the person to turn to one side agianst resistance applied by you to the chin. Repeat this on the other side. Observe the bulk of each sternomastoid and any weakness.

UMN lesions (supranuclear) of the accessory nerve cause mild weakness due to bilateral innervation. On the other hand, LMN lesions in the spinal cord cause true weakness of sternomastoid and trapezius. These lesions occur in:

- Amyotrophic lateral sclerosis
- Syringomyelia
- Poliomyelitis
- Tumours of the spinal cord

Wallenberg's Syndrome

This is caused by occlusion of vertebral or posterior inferior cerebella artery that produces infarction of the medullary tegmentum, and hence disfunction of the cranial nerves: fifth, ninth, tenth, and eleventh.

Tongue Muscles (CN XII)

Examine the tongue while resting on the floor of the mouth for atrophy and fasciculations. These are spontaneous contractions of muscle motor units caused by lower motor neuron disease like motor neuron disease.

To test for the twelfth nerve, ask the person tostick his tonue out, and see if there is any deviation. Also ask the person to move his tongue from side to side and test for weakness by trying to push it againstthe inner cheeks. LMN twelfth (hypoglossal) nerve palsy causes the tongue to deviate towards the weak side.

Diseases thatare weakness of the tongue:

- Muscles of the tongue
- Neuromuscular junction
- LMN lesion of hypoglossal nerve
- UMN cortex lesion, which causes contralateral weakness.

Cerebrovascular Accidents (CVAs)

This should not constitute a difficulty if a good history has been taken, anda perfect examination is performed.

Important points in the historyand examination:

1. Age of the patient and pre-existing diseases

Stroke in a young patient should alert you to take history of the onset, progression, associations with headache and palpitations and to look for:

a.   Cardiac disease: mitral stenosis, infective endocariditis, embolic events from metallic valves, or rarely atrial myxoma. Remember also patent foramen ovale.
b.   Connective tissue diseases and vasculitis (e.g., antiphospholipid syndrome, SLE).
c.   Vascular abnormalities: aneurysms in the brain.
d.   Secondary hypertension
e.   Use of the contraceptivepill

This is in contrast to CVA in middle age, or elderly patients in whom informative history and examination may reveal:

(1) Hypertension causing cerebral haemorrhage or infarction
(2) DM causing small vessel disease and infarction
(3) Atherosclerosis
(4) Carotid artery disease
(5) Cardiac disease in the form of: ischaemic heart disease and cardiomyopathy, causing emboli, especially in the presence of atrial fibrillation.
(6) Rarely, angoid amyloidosis

CVA Continued

(1) Remember that one of the causes of CVA is paradoxical emboli through a patient foramen ovale.
(2) As mentioned earlier, first give the anatomical diagnosis then the possible pathology and aetiology of such pathology.
(3) Logically differentiate between cerebral haemorrhage and infarction.

History	Haemorrhage	Infarction
Age	All ages, some young	All ages, more n >50
Risk factors	Hypertension A-V malformations Blood disease Anticoagulants	DM, Hypertension, cardiac disease
Sudden onset	Yes	Yes/No
Preceded by headache	Yes	Sometimes
Vomiting	Yes	Yes/No
Loss of consciousness	Yes/No	No

Examination	Haemorrhage	Infarction
Unconscious	Most	No/Yes (large infarction)
Neck rigidity	Yes, if extension to subarachnoid	No
Dense paralysis	Yes	Yes/No (large infarction)
BP	Very high/moderate	Normal/high
Pulse	Regular ±AF	AF/regular
Heart	Cardiomegaly/ normal	Cardiomegaly, murmurs
Carotid bruit	No	Maybe
Drugs (risks)	Antihypertensives Anticoagulant Antiplatelets	Antihypertensives

Difference Between <u>Cerebral</u> and <u>Brain Stem</u> Involvement

Higher function	Cerebral	Brain stem
Disturbance of conscious level	Yes/No	Yes/No
Vertigo	No	Yes
Speech	Aphasia	Dysarthria
Cranial nerves	UMNL seventh	Lower MN according to level
Hemiplegia	Uncrossed	Crossed
Sensory loss	Cortical pattern	Temperature, pain
Sympathetic involvement	No	Yes

Remember:

- Examine the eye movement for presence of homonymous hemianopia.
- Examine for sensory inattention
- Speech and type of aphasia

Other types of CVS (strokes):

(1) Anterior circulation with involvement of the parietal lobe. This usually gives ipsilateral hemiplegia with higher cerebral dysfunction and speech disturbance (sensory or motor). Paralysis may involve face or one limb only.

(2) Posterior circulation CVA (vertebrobasilar). In this type, you will find:

1. Ipsilateral paralysis (hemiplegia) with cerebral signs. Also diplopia and crossed hemiplegia.

Bilateral CVA

This will give bilateral UMN lesions of limbs together with pseudobulbar palsy. Look for:

(1) Dysphagia or nasal regurgitation
(2) Dysarthria
(3) Emotional disturbances: patient laughs or cries excessively.

Examination:

(1) Bilateral UMNL of limbs
(2) Spastic tongue
(3) Diminished movement of palate
(4) Brisk jaw jerk
(5) Remember, pseudobulbar palsy is also caused byMultiple Sclerosis (MS) and Motor Neuron Disease (MND).

Pseudobulbar vs Bulbar

	Pseudobulbar	**Bulbar**
Emotions	Labile	Normal
Speech	Slow/heavy	Nasal
Tongue	Small spastic	Fasciculations
Jaw jerk	Brisk	Normal/absent
Limbs	UMN lesions	LMN lesions
Anatomical site	Bilateral internal capsule	Medullar
Causes	Bilateral CVA, MS, MND	MND, polio, Guillain-Barre

Example of neurology cases(continued):

Paraplegia or quadriplegia

History: As mentioned before, take detailed history that include:

- The period before the onset of weakness: Any flu-like illness, headache, fever, backache, numbness, or visual disturbances. Ask about the mode of onset: sudden or gradual.
- The chronological from the start to the time for maximum weakness and indicate any occurrence of backache, root pain, or numbness.
- Sphincteric status: urine retention, faecal or urinary incontinence.
- Visual disturbance points towards multiple sclerosis, while cranial nerve system suggest Guillain-Barre syndrome.
- In case of quadriplegia, ask about arthritis (cervical spine involvement: atlantoaxial subluxation), tumour, and any sensory loss or burning of digits (syringomyelia)
- In patients more than forty years of age, ask about symptoms suggestive of a primary or secondary tumour.

Tips on how to perform CNS examination

(1) Attitude:

Wash hands, greet the patient by name if possible, introduce yourself, and explain the purpose and obtain permission.

(2) Tell him/her that you are going to examine the limbs and reassure him/her.
(3) Expose the limbs, but keep the patient well covered.

Patient position:

(1) According to the part to be examined and according to the comfort of the patient.
(2) **The upper limbs** are examined depending on the convenience of the patient. If the patient is able to sit at the side of the bed (if lower limbs are normal), this can be the position of choice. But if the patient has a neurological problem that also involves

the lower limbs, then the patient should be examined in the position adopted by the patient.

(3) Ideally, the whole limbs should be exposed in order to see tall muscles. However, it is important to respect patient culture, comfort, and dignity, and thus expose the limbs as far as an adequate examination can be performed.

(4) **When examining the lower limbs,** the situation is easier as the best position is the lying one with a pillow for convenience.

(5) **Exposure according to what is mentioned above**. Dresses should beremoved and the patient recovered, or rolled up to the groin.

Examination:

- **General:** Look at the patient as a whole and observe his general appearance (ill, distressed, drowsy), body built, position in bed, and pallor.
- **Go systematically:** Higher cerebral functions (level of consciousness, memory, orientation, and speech). Then examine the caranial nerves.

If asked in exams to to examine the limbs only, follow the following sequence:

- Inspect for abnormal position of limbs (externally of internally rotated, fixed flexion or extension).
- Also observe any wasting and where scars, ulcers, hyper or hypopigmentation, and fasiculations.
- Always inform the patient that his limbs will be examined, and ask him not to resist.

Upper Limb Examination

Test power, tone, coordination, reflexes, and sensations.

a) Examine power in all groups of muscles and compare the two sides.

b) Abduction of shoulder (deltoid C5). Ask the patient to spread his upper arms(chicken wings). Push inwards while the patient is pushing upwards and outwards. Adduction is the reverse of this action.

c) Flexion of the forearm (Biceps C5/6): ask the patient to make a fist and flexes arm at the elbow (towards him/her), while you pull or the flexed arm outwards.

d) Extension of forearm: Triceps (C7/8). The patient try to extend the flexed arm while you push against it.

e) Extension of he wrist (wrist extensors are supplied by the radial nerve): support the forearm with one hand, and ask the patient to make a fist facing downwards and push up. Your the other will resist the movement of the fist upwards.

Abduction of fingers(T1), Muscles: palmar interosseous innervated by theulnar nerve.

1. Ask the patient to hold his arm out with fingers spread. Then tell him/her to stop, you pushing them closed.

2. **Thumb abduction** (T1, median nerve). Ask the patient to raise the thum straight up while you push it down. Remember that the radial nerve supplies the extensors of the wrist, whilst the ulnar nerve supplies all of the intrinsic muscles of the hand, except for LOAF, which are supplied by the median nerve:

   Lateral two lumbricals (fingers two and three)
   Oppons pollicis
   Abductor pollicis brevis
   Flexor pollicis brevis

3. Testing for tone in upper limbs: passive flexion and extension of the wrists to find cogwheel rigidity of extrapyramidal diseases.

4. Then test for lead pipe and clasp-knife rigidity by flexing and extensing the elbow (put the two hand at the tendons). Also do supination and pronation of the forearm.

Coordination Cannot be Performed if there Is Significant Weakness

Upper limbs

Perform:

- Finger–nose test: detects intension tremor and past pointing. Instruct the patient to rapidlytouch the nose with the right index finger, and then touch the examiner finger, which will be moved to change position. Abnormal response is indicated by slow, clumsy response and past pointing.
- Dysdiadochokinesis: this is the rapid alternating movement of the one hand over the other. Abnormal response in cerebellar disorders. Ask the patient to hold the left hand palm out and the to clap with the right hand palm alternationg with back of the hand as rapidly as possible.

Lower Limb Examination

Inspection of upper limb:

1. Limb position: lateral rotation. Flexed, extended?
2. Skin: ulcers, pigmentation
3. Deformities
4. Muscle bulk: wasting, hypertrophy
5. Abnormal movement: Fasciculations
6. If possible, examine the gait: spastic (spastic paraplegia), hemiplegic, scissor's (inherited spastic problems), high steppage (peripheral neuritis), stapling (posterior colum), orfestinant (parkinsonism).
7. Test for Roberg's sign if suspecting peripheral neuritis.

While the patient is lying, test for tone.

- Instruct the patient to leave his limbs loose.

- Use both hands to roll over the extended limbs, and see if there is any resistance.
- Test for limb flopping by putting your hand below the knee and making an uward thrust.
- Then test for tone by trying to flex knee andankle. If you detect spasticity (upper motor neurone), test for pattellor and ankle clonus.

Testing for Power

Test all ranges of movements at the hip, knee, and ankle.

Hip power testing:

- Flexion (L1/2): Tell the patient to raise the leg with the knee extended. If this can be done, thenapply downward pressure at the quadriceps and judge the grade of power.
- Extension (L5/S1): Ask the patient to push the limb downwards while you resist this with your hand under the thigh.

Power at the Knee

a) Test for knee flexion (L5/S1): Ask the patient to flex the knees while you try to make it strait.

b) Knee extension (L3/4) is tested by asking the patient to do the opposite of knee flexion. Ask the parient to keep the flexed leg straight, while you are pushing against that. Tell the patient to kick out.

c) Ankle movement: Dorsiflexion (L4/5). Tell the patient to flex the ankle (push the toes up) while you prevent that.

d) Ankle plantar flexion (S1): the opposite of above. Ask the patient to push the foot down while you resist that.

Performing reflexes:

a) Knee reflex (L3/4): performed with the knees flexed. Place your left arm beneath the knees and tap the patellar. Report wether normal, depressed, or exaggerated. If exaggerated, do patellar clonus.

b) Ankle reflex (L5/S1): With the knee flexed and the hip externally rotated. Make sure the patient is relaxed and then tap the Achilles tendon. Observe for contraction at the calf and ankle.

Reinforcement:

This is used when there is no reponse from reflexes. Distract the patient by asking him to clench the teeth, or alternatively interlock the fingers of the two extended arms and ask the patient to pull them.

The Planter Reflex

(If there is extensor reflex and fanning of the toes, this is called Babiniski sign: L4-5 to S1-2)

a) Requirement: The patient should be calm and warm. Try to distract the patient, hold the resting foot from theankle while you stroke the lateral aspect of the foot. Start from the base of foot and go medially as you reach the toes.

b) What to use? Not a sharp object. Best is the neurology stick. Some doctors use car keys or the end of the tendon hammer.

c) The normal response is flexion of the big toe. If it goes up, then this extensor response(better not use the terms 'down going or upgoing'). If there is no obvious response, then report as 'equivocal' or 'no response'.

d) Some students and doctors confuse the planter reflex with the Babiniski sign. Only report that the Babiniski sign is positive if there is extension of the big toe and fanning of other toes.

Otherwise, only say the planter reflex is extensor or flexor (not upgoing or downgoing) if there is no fanning of the toes.

e)   Remember that an equivocal planter reflex is upnormal.

Alternatives to technique of eliciting the plantar reflex:

- **Chaddock's sign**: The stimulus is applied along thelateral aspect of the foot below the external malleolus.
- **Oppenheim reflex**: Firm pressure is applied alongthe shin of the tibia from below the knee upto theankle with the knuckles of the examiner's index and middle finger.
- **Gordon's sign**: The calf muscle is squeezed.
- **Schaefer's sign**: Squeezing the Achilles tendon.

History of the planter reflex and Babinski Sign

Joseph Babinski, a French neurologist, first differentiate between a normal and a pathologic plantar response, anddescribed the Babinski sign in 1896. The Babinski's sign isencountered in patients with pyramidal tract dysfunction, and is characterised by a dorsiflexion or extension of thegreat toe with or without fanning or abduction of theother toes. The fully developed response is alsoaccompanied by dorsiflexion of the ankle, flexion ofthe hip and knee joint, and slight abduction of the thigh, leading to a withdrawal of the leg on plantar stimulation.

The extensor plantar reflex is caused by a umber of conditions headed by the pyramidal tract lesions

Other causes:

1.   Children up to a year old
2.   Coma
3.   General anaesthesia
4.   Deep sleep

5. Hypoglycaemia
6. Heavy alcohol intake
7. Narcosis
8. Hypnosis
9. Postictal state of epilepsy
10. Electroconvulsive Therapy (ECT)
11. Afer heavy physical exhausion

Sensation:

(1) Test all modalities, as for upper limb. It may be worth performing Romberg's test whilst the patient is standing at the start of the examination.
(2) Use small thin cotton fibre for light touch.
(3) Pain and temperature: use neurology pin.
(4) Posterior column sensations: vibration, position, and proprioception.

(Includes Romberg's sign)

Romberg's Test is determinator of cerebellar (unsteady all the time) versus proprioceptive (peripheral) sensory input deficit (in which case, the patient will be increasingly unsteady with eyes closed).

Ask the patient to stand with legs next to one another and arms by side. Then ask them to close their eyes. Make sure you support them if they are unsteady. A positive test is when they are more unsteady when their eyes are closed indicates a peripheral sensory problem.

Remember:

Find if there is a clear sensory level, and what sensations are lost or diminished.

(5) Paraplegia without sensory loss occurs inparasagital lesions(e.g., meningioma). Also, in motor neuron disease, in this case, there

is wasting of small muscles of hands, fasciculations, and absent reflexes in upper limbs and pyramidal signs in the lower limbs. Also, there are bulbar signs. However, cervical cord compression may give a similar picture.

Grading of the limb weakness (out of 5):

- 0: No movement at all. Complete paralysis
- 1: Flicker of movement
- 2: Movement without effect of gravity
- 3: Movement against gravity, but not resistance
- 4.: Moderate movement against resistance
- 5: Normal power

Differences between upper and lower motor neurone legions:

Feature	UMN	LMN
Muscle wasting	Little/None	Present
Tone	Hypertonia	Hypotonia
Fasiculations	None	Present/none
Weakness	Upper limb extensors and lower limb flexoers	Global weakness
Reflexes	Brisk	Diminished/absent
Clonus	May bepresent	Absent
Planter reflex	Extensor/Babinski	Flexer
Sensations	Sensory level/cortical	Depends on cause

Remember: In acute stage of upper neurone lesions, tone and reflexes may be diminished due to neuron shock.

Other Short Neurological Cases

(1) Brown-Séquard syndrome (hemisection of the cord) Brown-Séquard-plus syndrome

The pure Brown-Séquard syndrome occurs more in the cervical cord, reflecting hemisection of the cord is not often observed. A clinical picture composed of parts of the syndrome or of the hemisection syndrome, plus additional symptoms, and signs is more common. These lesspure forms of the disorder are often referred to as Brown-Séquard-plus syndrome.

Interruption of the lateral corticospinal tracts, the lateral spinal thalamic tract, and at times, the posterior columns produces a picture of an ipsilateralspastic, weak leg with brisk reflexes, and a contralateral normal leg with loss of pain and temperature sensation. Note that spasticity and hyperactive reflexes may not be present with an acute lesion.

Findings below the lesion:

Same side of the lesion

(1) Monoplegia or hemiplegia
(2) Loss of joint position and vibration sense

Opposite side

- Affection of the spinothalamic tract (loss of pain and temperature)findings in the segment of the lesion

(3) Same side lower motor neurone paralysis
(4) Same side zone of cutaneous anaesthesia

Causes:

(5) Cord tumour
(6) Haematomyelia

(7) Syringomyelia

(8) Degenerative diseases

(9) Trauma

Short Neurological Cases

(1) <u>Multiple sclerosis</u>
<u>Read the following story:</u>

Ms A. is a 30-year-old female. The paient reports that during the last two years, she noticed some clumsiness anddisturbed gait so that she often stumbles. She also reports that she experience some visual abnormalities. Futhermore, she reports that whenever she gets flu or she is under tension and stress, the neurological problem get worse.

For the last six weeks, her coordination got worsewith hand tremor, so thatshe occasionally drops spoons and cups. She also described what seemed to be left hemisensory changes in her body.

The patient also reports that she had bladder disturbances requiring frequent visits to the toilet, and lately, she became incontinent and needed to wear a pad.

When seen at that time, an MRI revealed white matter abnormalitieswith intense T2 signals in both hemispheres. Further evaluation showed abnormal visual evoked reponses with slow conduction in optic nerves. A lumber pucture and CSF examination revealed an oligoclonal band.

Q1: List all the symptoms in this patient's history that may lead you to suspect the diagnosis of multiple sclerosis. List all the symptoms and findings in the order of their importance.

Answer:

1. Numerous neurologic complaints, which came and went over a number of years and affected several different neurologic systems.
2. Generalised white matter disease as documented by MRI imaging.
3. Presence of oligoclonal bands in CSF.
4. Abnormal visual evoked responses with slowed conduction in the optic nerves.
5. The patient is a 30-year-old white female.
6. Some neurologic symptoms were precipitated by heat intolerance.

Neurologic examination:

Eye examination (cranial nerve II) showed optic neuritis. **Examination for eye movements (Cranial Nerves III, IV, VI)** did not show any ocular muscle weakness.

The rest ofof the cranial nerve exam is normal except for decreased hearing on the left and numbness in the right face, which extends down into the entire right side. The Weber test reveals greater conductance to the right. Rinne test reveals air greater than bone bilaterally. The palate elevates well. Swallow appears to be intact. Tongue movements are slowed, but tongue power appears to be intact.

Limb examination for motor function, there isnormal strength in the upper extremities throughout. However, rapid alternating movements are decreased in both upper extremities, and the patient has dysdiadochokinesia in the left hand. Mild paraparesis is noted in both legs without severe spasticity.

Deep tendon reflexes are +2 and symmetrical in the arms, +3 at the ankles and at the knees. Bilateral extensor toe sign are present. Sensory exam reveals paresthesia on the right to touch, and decreased pin sensation on the right diffusely. The patient has mild vibratory sense loss in the distal lower extremities.

Romberg's is negative. Tandem gait is mildly unstable. Ambulation index is 7.0 seconds for twenty-five feet (the patient takes 7.0 seconds to walk 25 feet).

Q1: Based on the clinical findings, what CNS pathways are affected?

Answer: The clinical findingsindicate dysfunction of thecerebellar pathways. This can be assessed further by performinga 'heeltoshin' test to check for coordination abnormalities in lower extremities and 'fingertonose' test for upper extremities. As well, testing for tandom gait is another test for cerebellar functions.

Q2: What is the best way to perform Romberg sign?

Answer: Have the patient stand withfeet together, arms stretched in front. Then ask the person to closeeyes and watch the balance. Romberg is a test of the dorsal column—medial lemniscus function.

'Negative Romberg test' refers to a stable, well-balancing patient witheyes either open or closed. Romberg test is considered positive if the patient stands on a narrow base with eyes open, but falls on closing the eyes.

In MS, we may see posterior column findings, cerebellar findings, or both contributing to loss of balance. Romberg test helps to assess posterior column dysfunction, but detailed sensory and cerebellar exam may be required to discern the causes of imbalance. Romberg test is not very informative if the patient's motor strength in the lower extremities is severely compromised.

Q3: What other conditions can result in generalised white matter disease evident on the MRI?

Answer: Cerebrovascular disease (i.e., 'small strokes'), HIV encephalitis, progressive multifocal leukoencephalitis, systemic lupus erythematosus, neurosarcoidosis, spinocerebellar degeneration, BehTauet Syndrome, normal ageing.

Usually, you are asked to examine the legs of the patient who is lying in bed or in a wheelchair.

A thorough physical examination, including neurologic assessment, is critical to determine deficits in MS. All systems must be addressed including cognition, mood, motor, sensory, and musculoskeletal, as well as the following:

- Reflexes
- Coordination
- Bulbar function
- Vision
- Gait
- Skin

Bulbar involvement typically refers to dysfunction of lower cranial nerves whose nuclei reside in the lower brainstem. Manifestations include dysphagia, which does not occur often in early MS, and so may be attributed to a different disorder.

Patients with MS may demonstrate a variety of abnormal physical findings, and these findings may change from examination to examination, depending on the pattern of disease, and whether the patient is having an exacerbation or relapse. Findings may include the following:

- Localised weakness
- Focal sensory disturbances (with persistent decrease of proprioception and vibration)

- Hypereactivereflexes with clonus in the ankles and extensor planter reflex.
- Increased tone or stiffness in the extremities with velocity-dependent passive range of motion.

Additional signs may include poor coordination of upper and lower extremity movements, the Lhermitte's sign, and wide-based gait with inability to tandem walk.

Observe the general look of the patient and proceed to examine the lower limbs as usual.

(2) <u>Expected findings in lower limbs</u>

Spastic paraparesis.

(1) Impaired coordination (difficult in the presence of spasiticity).
(2) No definite sensory level.
(3) Absent or diminished abdominal reflexes.

Saythat you would like to examine for cerebellar signs and look for optic atrophy.

Suggested investigations:

- Lumber puncture: raised CSF proteins and increased immunoglobulin G (40%) and oligoclonal bands on electrophoresis (80%).
- Visual evoked potentials: increased latency in cortical response.
- MRI (T2—weighted) abnormal signals (hypertense) in periventricular white matter of cord (50%) and brain (80%).

Criteria for Categorizing MS

MS is divided into the following categories, principally on the basis of clinical criteria, including the frequency of clinical relapses, time to disease progression, and lesion development on MRI.

- Relapsing-remitting MS (RRMS)
- Secondary progressive MS (SPMS)
- Primary progressive MS (PPMS)
- Progressive-relapsing MS (PRMS)

RRMS is characterised by recurrent attacks in which neurologic deficits appear in different parts of the nervous system and resolve completely, or almost completely, over a short period of time, leaving little residual deficit. Patients with a relapsing-remitting pattern account for approximately 85 per cent of MS cases (see the images below).

Management

(1) General
(2) Methylprednisolone
(3) Interferon Beta
(4) IV Immunoglobulins
(5) Plasma exchanges
(6) Bladder: increased residual urine
(7) Volume: oxybutinine

Emergency Department Management

Medical management goals are sometimes achievable in the emergency department are to relieve symptoms and to ameliorate risk factors associated with an acute exacerbation. In patients with fulminant MS or disseminating acute encephalitis, management involves the following:

- Stabilise acute life-threatening conditions
- Initiate supportive care and seizure precautions
- Monitor for increasing intracranial pressure

Consider intravenous steroids, IV immunoglobulin (IVIG), or emergent plasmapheresis. One study suggested that plasmapheresis may be superior to IV steroids in patients with acute fulminant MS. The 2011 American Academy of Neurology (AAN) plasmapheresis guideline update states that plasmapheresis is possibly effective, and may be considered in acute fulminant demyelinating CNS disease.

Identification and control of known precipitants of MS exacerbation include the following:

- Aggressively treat infections with antibiotics
- In patients with a fever, normalize the body temperature with antipyretics, as even small increases in temperature can strongly affect conduction through partially demyelinated fibres.
- Provide urinary drainage and skin care, as appropriate.

Some of the emerging therapies, including alemtuzumab, daclizumab, rituximab, ocrelizumab, laquinimod, estriol, 3-hydroxy-3-methylglutaryl-coenzyme A (HMG-CoA) reductase inhibitors (statins), vitamin D, and stem cell transplantation, what neurological problems that you can get when having a hair cut?

Barber's Chair Sign or Lhermitte's Sign

Electric shock-like (tingling sensation), which pass down the arms, back, or legs on flexion of the patient's neck.

Occur in multiple sclerosis: due to cervical dorsal column nucleiinvolvement.

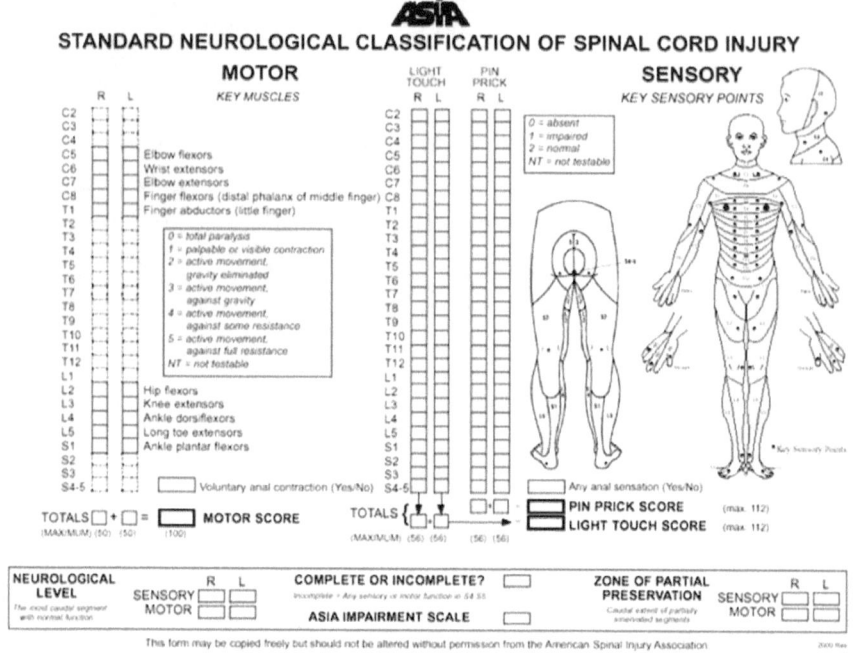

Fig 18: Summary of spinal cord lesions

More Neurology Examples

o **This patient has weakness of the limbs. Examine the CNS.**

The patient was a 73-year-old lady who was seated comfortably. She was of average height and weight. On examination of the upper limbs, there was no major change in size or shape of the limbs. The skin was normal.

On examination of the muscles, they noticed that there was some wasting of the small muscles of the hand involving the thenar eminence bilaterally and the interossei. However, the wasting was not uniform. She also noted fasiculations in the triceps muscles.

On examination of power, there was weakness mostly distally (small muscles of the hand) the distribution of the weakness was not uniform.

Reflexes were exaggerated bilaterally and Hoffman's sign was positive. Exaggerated reflexes were evidence that the upper motor neurons were involved, and this suggested that the patient had amyotophic lateral sclerosis.

There were no overt cerebellar signs, and there was no sensory deficit. This was further evidence in favour of motor neuron disease.

Diagnosis: Motor neuron disease with amyotrophic lateral sclerosis

o **Examine this patient with difficulty using his hands.**

General:

Patient is sitting comfortably. He is of average built and not connected to any device.

No facial abnormality observed. Eye movement and facial muscles are normal.

Upper limbsexamination and inspection:

Shape, size, and skin are normal. In particular, no wasting or fasiculations or abnormal movement.

Tone: Normal at wrist and elbow, but flaccid at the shoulder as the arms fell when raised above the head. This indicates lower motor neuron disase or myopathy.

Power: Starting from the small muscles of the hands and proceeding upwards, there was G2 weakness of abduction and adduction of the shoulders. Also, there was wasting of muscles around the scapula.

What does this suggest? Possibly lower motor neuron at C4, C5, or myopathy.

Coordination: poor due to muscle weakness.

Reflexes: Brisk reflexes of the biceps and triceps. Also, an inverted supinator reflex and positive Hoffman's sign. What do these findings mean? They indicate a lower motor neuron lesion at C4-C5 together with a lower motor neurone below that level.

What could be the cause of these findings?**Most likely cervical myelopathy.**

What pattern of sensory loss is expected?In fact, when sesations are examined, there was sensory impairment along C4-G5 ermatome in the hands.

What is the final diagnosis?This patient has cervical myelopathy affecting C4-C5 roots. Besides blood workup, the patient needs MRI of the cervical spine. This may show disc compression or inflammatory conditions. Another possibility is a mass legion.

Practical hints:

- **Start examination by inspection, so that you do not miss wasting especially here around the scapula.**
- **Examine patients systematically. Inspection, tone, power, coordination, reflexes, and sensory assessment.**
- **Correct interpretation of findings. Here, the inverted supinator reflex gave important information. Revise the proper techniques of eliciting neurological findings.**
- **Practise eliciting special tests like invertion of reflexes and Hoffman's sign.**

Sensory loss and corresponding nerve root:

1. Posterior aspect of the shoulders (C4)
2. Lateral aspect of the upper arms (C5)

3. Medial aspect of the lower arms (T1)
4. Tip of the thumb (C6)
5. Tip of the middle finger (C7)
6. Tip of the pinky finger (C8)
7. Thorax, nipple level (T5)
8. Thorax, umbilical level (T10)
9. Upper part of the (knees) leg (L2)
10. Medial part of the upper leg (L3)
11. Medial lower leg (L4)
12. Lateral lower leg (L5)
13. Sole of foot (S1)

o **Examine this patient who complained of weak legs.**

General examination: a young patient, average weight, no distress or abnormal posture.

On examination of the lower limbs, there was noskin changes and no difference in bulk of the limbs. This indicates that the problem is acute. Obvious wasting is noted without any abnormal movement or fasciculations. This indicates that we are dealing with a LMN lesion without anterior horn cells deficit.

Tone: flaccid, which proves further a LMN lesion.

Power: G3, especially the distal muscles. Thus, we have a patient with flaccid paraparesis.

Reflexes: absent reflexes, indicating a lowermotor neurone lesion.

Senasations: mild loss of pain and touch over feet. No specific pattern of cauda equina lesion that is saddle loss. Also, no definite sensory level to help localise the site of the lesion.

What is the differential diagnosis?

A lower motor lesion of this type is caused by radiculoneuropathy or a cauda equina lesion. In this patient, a polyradiculoneuropathy is most likely. This can be due to inflammatory or multiple compression.

What further question would you ask the patient?

One should ask about sphincter control especially bladder and rectum. The bladder and rectum are commonly affected in cauda equina syndrome rather than in inflammatory polyradiculoneuropathy such as Guillain-Barre syndrome.

Also, one would like to do rectal examination looking for anal tone and reflex, which were typically reduced in cauda equina lesions.

Complete Diagnosis

This patient is suffering from acute flaccid paraparesis due to a possible inflammatory demyelinating polyradiculoneuropathy process such as Guillain-Barre syndrome.

Differential Diagnoses of GBS:

- Botulism
- Cauda Equina and Conus Medullaris Syndromes
- Chronic Inflammatory Demyelinating Polyradiculoneuropathy
- Emergent Management of Myasthenia Gravis
- Heavy Metal Toxicity
- Lyme Disease
- Metabolic Myopathies
- Multiple Sclerosis
- Nutritional Neuropathy
- Vasculitic Neuropathy

What Investigations Would You Suggest?

1. Electromyography (EMG) and Nerve Conduction Studies (NCS) can be very helpful in the diagnosis. Abnormalities in NCS that are consistent with demyelination are sensitive and represent specific findings for classic GBS. Delayed distal latencies, slowed nerve conduction velocities, temporal dispersion of waveforms, conduction block, prolonged or absent *F* waves, and prolonged or absent H-reflexes are all findings that support demyelination.

2. Lumbar puncture for Cerebrospinal Fluid (CSF) studies is recommended. During the acute phase of GBS, characteristic findings on CSF analysis include albuminocytologic dissociation, which is an elevation in CSF protein (>0.55 g/L) without an elevation in white blood cells. The increase in CSF protein is thought to reflect the widespread inflammation of the nerve roots.

3. Imaging studies, such as Magnetic Resonance Imaging (MRI) and computed tomography (CT) scanning of the spine, may be more helpful in excluding other diagnoses, such as mechanical causes of myelopathy, than in assisting in the diagnosis of GBS.

4. Frequent evaluations of pulmonary function parameters should be performed at bedside to monitor respiratory status and the need for ventilatory assistance.

o **Other Difficult Neuro Cases**

• **Cerebellar problems**

This patient presented with unsteady gait

Examine for cerebellar signs:

• Test for Dysdiadochokinesia by asking the patient to perform rapid alternating movement

- Perform the heel–shin test to look for ataxia
- Examine eyes for nystagmus
- Do the finger–nose test looking for intension tremor
- Observe the slurred speech called staccato speech
- Ask the patient to do heel-toe walking

Remember cerebellar signs by DANISH:

Dysdiadochokinesia
Ataxia (lower limbs, impaired heel–shin)
Nystagmus (ipsilateral)
Intention tremor (on finger–nose test)
Staccato speech (slurred)
Heel-toe walking impaired

- **Charcot-Marie-Tooth Disease**

Examine this patient who complains of leg numbness.

- Look for typical distal wasting, which looks likeinverted champagne bottle
- Observe pes cavus
- Test for peripheral neuropaty (mixed motor and sensory)
- Observe the high stepping gait

Mention some types of Charcot Marie Tooth

There are five types: axonal, demyelinating, and many genotypes. It is inherited as AD affecting 1/2500.

What is the natural history?

- Affectsteens, twenties
- May cause severe neuropatic syndromes

What further tests are needed?

Nerve conduction studies and genetic testing are helpful. Usually, there is prolonged conduction together with demyelination and decreased amplitiude of axonal APs.

What other conditions are associated with Pes cavus?

- Charcot Marie Tooth (HMSN)
- Freidrich's Ataxia
- Idiopathic - 20 per cent
- Fracture malunion
- Secondary to contracture
- Polio
- Syringomyelia

- **A patient with Myotonic dystrophy**

Inherited as Trinucleotide repeatAD

What to look for on examination?

Look first:

- Dull looking face: myopathic facies
- Wasting of face, neck, and distal arms
- Frontal balding

Examine for

- Shake hands to confirm hand-grip myotonia
- Also, elcit percussion myotonia
- Eyes for cataracts

- Confirm decreasedpower in face and bulbar muscles
- Observe the slurred nasal speech
- Test to elicit distal weakness
- Perfom the test for reflexes. Usually, there are abscent reflexes.

What further steps to conclude the examination:

- I would like to test for cognitive function
- Examine the CVS for cardiomyopathy
- Examine genitalia for hypogonadism (testicular atrophy)
- Urine dip for glucose – they get diabetes

Investigation that can help

- Serum creatinine kinase
- Electromyography
- Genetic studies for MD Type 1
- Lung function tests
- ECG and Echo
- Workup for DM
- Formal cognitive functions

What happens long-term?

The condition is progressive and weakness willlead to disappearance of myotonia.

Mention few other myopathies:

- Duchenne's and Becker's (childhood)
- Facioscapulohumeral and limb-girdle muscular dystrophy (adulthood)

Further Difficult CNS Problems

- **Examine this patient presenting with numb feet.**

What to do?

- Inspect the patient for any neurological abnormalities such as myopathic face or abnormal position. Inspect the skin for trophic ulcers or scars.
- Examine systematically the motor system for power. Tone, coordination, and reflexes.
- Note that the Achilles reflex is the first to be lost because of the large myelinated fibres.
- Examine sensations looking for the typical glove and stocking distribution
- Test for vibration, which is lost early (dorsal column)
- Then test for propioception (dorsal column). A fibre, large myelinated.
- Lastly, for pain and temperature (spinothalamic tract). C fibres, which are thin unmyelinated.
- You may end by testing the autonomic system (B fibre intermediate).

Discussion: Mention the clinical diagnosis. Is it motor, sensory, or both? Then suggest the possible causes accordingly.

Investigations

- CBCand ESR: anaemia and type
- Biochemistry, especially blood sugar, Ca, phosphorous, and alkaline Phosphatase
- Assement for DM HbA1C
- Folic acid andB12
- Workup for the thyroid
- Immunoglobulins, HIV and lyme disease workup
- Rarely porphyria workup

Incase you found mixed motor and sensory, look for:

- Pes cavus –Charcot Marrie Tooth(CMT)
- Diabetes – ulcers, injection sites, finger-prick marks
- Rheumatoid (Amyloid)

Some Causes:

Acute

- Guillain-Barre - motor
- Diphtheria - mixed
- Porphyria - motor

Causes of subacute neuropathy:

Drugs: mainly mixed (predominantly sensory)
Isoniazid, Metronidazole, Nitrofurantoin
Dapsone: motor
Cisplatin, Vincristine
HIV medications
Alcohol
Toxins
Lead - motor
Solvents - mixed
Industrial compounds
Nutritional
B12 - painful sensory with severe later motor symptoms

Chronic neuropathy:

Paraneoplastic - sensory
Paraproteinaemia - mixed
CTD - mixed
Amyloid - mixed, with autonomic features and entrapments

Uraemia - mixed
Hypothyroidism - mixed
Diabetes - mixed
CIDP
Hereditary
Refsum's Dx

Causes by group:

- **Metabolic** – hypothyroidism, B12 and diabetes, Porphyria
- **Drugs** – alcohol, chemotherapy, HIV medications, antibiotics
- **Paraneoplastic** (anti Hu)
- **Inflammatory** – SLE, RA, Guillain-Barre, Amyloid
- **Infectious** – HIV, syphilis, lyme, leprosy
- **Hereditary** – CMT and Refsum's
- **Heavy metals and solvents**

Pure motor:

a. Guillain-Barre
b. Diphtheria
c. Porphyria
d. Lead
e. Dapsone

The Cerebellar Syndromes

Somecauses: vascular, degerative, neoplastic, and alcohol

Examination

Face

- Look for nystagmus – towards lesion
- Observe speech – slurring, Staccato
- Titubation

233

In the arms, test for:

- Rebound with eyes closed
- Dysdiadochokinesia: performing alternating movement
- Intention tremor
- Dysmetria - past pointing

In legs, look for:

- Motor power: less
- Tone: hypotonia
- Coordination: heel–shin
- Reflexes: pendular
- Giat: ataxic

What are the causes of bilateral cerebellar syndrome?

Congenital:

- Agenesis
- Dandy-Walker: congenital malformation involving the cerebellum with enlargment of the fourth ventricle.
- Arnold-Chiari: underdevelopment of cerebellum with herniation down the foramen magnum. Three types.
- Von Hippel-Lindau: inherited disease with tumour and fluid spaces(haemangioblastoma)

Neoplastic:

- Primary or secondary
- Paraneoplastic: especially small cell lung cancer, lymphomas, ovarian

Degenerative:

- Spinocerebellar ataxia

- Multi-system atrophy
- Cerebellar ataxia
- Cerebellar degeneration
- Friedreich's ataxia

Other signs of Friedrich's:

- Kyphoscoliosis
- Optic atrophy
- Diabetes
- Sensory neuropathy
- Cardiomyopathy
- Genetic abnormality – AR – Trinucleotide repeat - anticipation

Metabolic Diseases:

- Thiamine deficiency
- Hypothyroidism
- Hypoglycaemia

Medications:

- Phenytoin
- Ethanol
- Lead

Vascular:

- Cerebrovascular accidents
- Haemorrhagic

Inflammatory/Infection:

- Viral
- Abscess formation

- **Internuclear Ophthalmoplegia (INO)**

  - Conjugate lateral gaze disorder
  - Failure of adduction on lateral gaze, often with nystagmus
  - Due to a deficit in the coordination of vision in the medial longitudinalfasciculus in the pons (between the CN VI and CN III nuclei)

Causes of INO

If unilateral: think of stroke, MS, and trauma

Bilateral causes: syringomyelia, MS, Phenytoin toxicity, brainstem disorders, and trauma.

What is One and a Half(½) Syndrome?

In this rare disorder, there's a failure of lateral gaze palsy in one eye and internuclear ophthalmoplegia in the other.

It is caused bya unilateral lesion of the dorsal pontine tegmentum, which involves the ipsilateral paramedian pontine reticular formation, internuclear fibres of the ipsilateral medical longitudinal fasciculus, and usually the abducens nucleus.

Similar to INO, but involving the PPRF and the MLF, which causes failure of anylateral movement of the affected eye and failure of adduction on the contralateral eye, leaving only abduction of the other eye still present. The main causes are MS and stroke.

- **Myasthenia Gravis**

First, inspect for the myopathic faceand partial ptosis. Look for anythymectomy scar, then do the following examination:

1. Inducefatigability by asking the patient to look up and count to fifty. Ptosis and neutral gaze will then be evident.
2. Ask the patient to smile, and observe the buccinator weakness with myasthenic snarl.
3. Observe the nasal speech.
4. Test for proximal weakness by asking the patient to raise his/her hands upthen look for distal weakness.
5. Thesensation and reflexes are usually normal

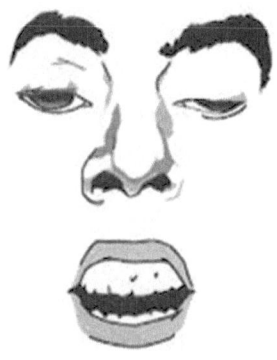

Fig: 19: Myathemia Gravis

Eyes

Look for:

- Movements (may have opthalmoplegia) and diplopia
- Fatigability(test-sustained upgaze)
- Fatiguability(test repeated blinking)

Speech

Observe:

- Nasal sounding speech(bulbar-like palsy)
- Get the patient to count upwards(may slow down/change character of speech as they fatigue).

Arms

Test for:

- Ask the person to make chicken wings type movements (to demonstrate fatigability)
- Be careful not to cause overfatigue to the patient.

Note about Myasthenia:

Myasthenia gravis is a neuromuscular junction disorder characterised by weakness ofsome muscle groups with fatigability. It is caused by autoimmune destruction ofpostsynaptic nicotinic receptors by acetylcholine receptor antibodies (complement mediated). Thymic is also involved.

Suggested investigations:

- Acetylcholine receptorAbs
- Edrephonium (Tensilon)test
- Electromyography
- CT Chestto detectthymoma

Management:

- Give Anticholinesterase drugssuch as pyridostigmine, pyridostigmine that is longer, acting with half-life of four hours.
- Also, atropine to reduce the muscarinic effects
- Steroids
- Steroid-sparing agents
- Consider thymectomy (if thymoma or v severe)
- IVimmunoglobulinesand plasmaphoresis

Complications

Myasthenic Crisis

Management:

- Manage airway
- Oxygen and monitor CO2
- IV neostigmine
- Subcutanousatropine
- High-dose prednisolone
- Consider plasmapheresis or IV immunoglobulin

Other complications:

- Differentiate between cholinergic andmyasthenic crisis
- Respiratoryfailure
- Aspiration
- Neonatal myasthenia from AB crossing placenta

Quick Brief Neuro Problems

A patient with absent ankle jerks and extensor plantars. What could be the cause?

Causes:

Subacute combined degeneration cord
Syphilitic taboparesis
Friedreich's ataxia
Motor neuron disease
Two othercommon conditions (e.g., DM + cervical spine compression)
Lesion of conus medullaris

Notes

- Friedreich's ataxia: look for pes cavus, kyphoscoliosis, cerebellar signs, impaired joint position/vibration, cardiomyopathy, optic atrophy, diabetes mellitus, mild dementia
- Severe Vitamin B12 defiency: subacute combined degeneration (peripheral neuropathy, pos column signs, pyramidal signs)

Bulbar/Pseudobulbar Palsy

Causes of bulbarpalsy:

- Patients with motor neurone disease
- Observe nasal speech, tongue fasciculations, and absent jaw jerk.

Differential diagnosis of bulbar versus pseudobulber palsy:

Bulbar	Pseudobulbar
MND	Bilateral CVA internal capsule
Syringobulbia	MND
Guillain-Barre	Multiple sclerosis
Polio	

Other RareNeurologyProblems

Hereditary Motor Sensory Neuropathies (HSMN)

Charcot-Marie-Tooth disease (CMT1 = HMSN1) or peroneal muscular atrophy is the most common example in clinical practise.

Main signs: distal leg wasting, pes cavus, foot drop, absent ankle reflex, and distal sensory loss.
Assess disability and function.
Differential diagnosis of HSMN is Friedreich's ataxia.

Further tests to do:

- Examine gait(high steppage or foot drop)
- Feel for lateral popliteal nerve thickening
- Look for small muscle wasting hands

Notes

- Cardiac manifestations include both aortic regurgitation and AD modes of inheritance.
- Type I = demyelinating, absent reflexes; II = axonal, normal reflexes

Characteristics of the main inherited neurological disorders:

Main clinical features	Frequency*	Age of onset	Genetics	Disorder
Neurofibroma on nerves, café-au-lait spots >5, axillary freckling	1/4,000	Children and adults	Autosomal dominant	**Neurofibro matosis type 1**
Deafness, bilateral acoustic neuromas	1/50,000	Adults	Autosomal dominant	**Neurofibro matosis type 2**
Epilepsy, adenoma sebaceum on face, hamartoma	1/15,000	Children and young adults	Autosomal dominant	**Tuberous Sclerosis**
Cerebellar signs, progressive gait ataxia, upper motor neurone signs, absent ankle jerks	1/50,000	Teens	Autosomal recessive	**Friedreich's ataxia**

| Peripheral neuropathy, marked lower limb distal wasting, high arched feet (pes cavus) | 1/3,000 | Young adults | Autosomal dominant | **Charcot-Marie-Tooth disease** |
| Choreaform movements, progressive dementia, psychiatric symptoms | 1/10,000 | Middle-aged adults | Autosomal dominant | **Huntington's disease** |

Examine this patient with limb weakness: motor neurone disease.

    a. Observe: is the patient on wheel chair? (Tracheostomy)
    b. Note speech problem: heavy slurred
    c. Tongue and limb fasciculations
    d. Upper limbs: wasting of small muscles and weakness
    e. Brisk jaw reflex
    f. Lower limbs: upper MN signs

Differential diagnosis: cervical cord compression, old polio, spinal muscular atrophy of juvenile onset type three.

Tests: Needs MRI cervical spine

Motor Neurone Disease

Important clinical notes:

- Key feature is combination of upper and lower motor neuron signs. No occular involvement, and no cerebellar signs.

Disease patterns: (i) Bulbar: bulbar/pseudobulbar palsy; (ii) amyotrophic lateral sclerosis: flaccid arms, spastic legs; (iii) progressive muscular atrophy: distal muscles; (iv) primary lateral sclerosis: progresses from UMN to LMN type.

Drug therapy: Riluzole (glutamate antagonist) has some evidencein patients with bulbar onset with lower risk of death or tracheostomy. However, it isnot clear if this translates to improved quality of life.

A Patient with Multiple Sclerosis

Patient either in wheel chair, paraplegic, or walking with difficuly:

a. Note the clumsy hands(disturbed coordination)
b. Optic atrophy (young females), internuclear ophthalmoplegia
c. Crebellar signs: nystagmus, disturbed coordination, and rapid alternating hand movements
d. Spastic paraparesis
e. What tests: Vitamin B12 to exclude subacute degeneration

   MRI brain: hyperintense focal perventricular white matter lesions on T2 weighed

   Lumber puncture:↑Total protein, oligoclonal bands

   Serum electrophoresis(to exclude oligoclonal bands in serum)

Types:

- Relapsing-remitting, secondary progressive, primary progressive, progressive relapsing
- Poor prognosis: progressive disease, multiple lesions on MR, frequent relapses first two years
- IV methylpred is given to speed recovery in relapses. However, no evidence for long-term benefit.

- Interferon beta (1a/1b) reduces the relapse rate in relapsing/remitting by one-third. Indicated for ambulant patients with at least two relapses in previous two years with recovery.

Patient with Myotonic Dystrophy

Note Myopathic face:

1. Ptosis, wasting tempolaris, cataract
2. Frontal baldness (wearing wig?)
3. Shake hands (myotonia)
4. Depressed reflexes

What tests to do?

1. Blood sugar fordiabetes mellitus
2. ECG (conduction abnormalities)
3. EMG (waxing/waning of potentials 'dive-bomber'effect)
4. Slit lamp examination (cataracts)

Things to do:

- Ask patient to make a fist then open hand quickly
- Open eyes quickly after firm closure
- Percussion myotonia(dimples in muscles fill slowly)

Notes:

- Other features: cardiomyopathy, conduction defects, respiratory infections (low IgG levels)

Differential diagnosis:

- Other causes bilateral ptosis: Myasthenia gravis, muscular dystrophies

- Myotonia congenita (Thomsen's disease): myotonia but no other features of MD, reflexes normal, can have Herculean appearance

Patient with Parkinson's Disease

- Patient with speech problem: scanning heavy speech
- General: expressional mask face, heavy speech, drooling saliva
  Gait: shuffling andfestinating, absence of arm swing

  Tremor: pill-rolling 3-5Hz tremor

- Bradykinesia and generalised rigidity(mainly cogwheel)

Aetiology ofParkinsonism:

- Primary, degenerative
- Brain anoxia: CO
- Postencephalitis
- Drugs: phenothiazines

More clinical signs:

- Test gait
- Glabellar tap
- Check for righting reflex (micrographia)
- Tremor: Emphasise by mental distraction (ask patient to count backwards from twenty)
- Test for rigidity: assess upper limb rigidity and ask the patient to move the other limb up and down at the same time.
- Bradykinesia: ask patient to open and close hand quickly.
- Function(difficult tasks): ask the patient to drink water 'undo button'.

Assess for other Parkinsonian conditions:

- Check for autonomic dysfunction: lying, standing BP (multi-system atrophy)
- Eye movement, particularly upwards gaze (supranuclear palsy)
- Cognitive function (lewy body dementia)

Management:

- L-DOPA with decarboxylase inhibitor (carbidopa/benserazide)
- Ropenirole (D2 agonist) reduces risk of dyskinesia, but not on/off. Selegiline (MAO B inhibitor) delays need for levodopa in early disease
- Deep brain stimulation of globus pallidus and subthalamic nucleus

A Patient with Peripheral Neuropathy

a. Presentation according to type: motor, sensory, and motor-sensory
b. Presence of a cause: DM, neutritional, malabsorption, and connective tissue disease. Ask about ethanol, B12 defiency, and drug therapy (metronidazole).
c. Findings: gloves and stockings sensory or motor loss.

Common Causes	Pure Motor
Diabetes mellitus	Lead
Alcohol	Porphyria
Vitamin deficiency(e.g., B12)	Diphtheria
Drugs: isonizid, vincristine	
Inflammatory: RA	
Paraneoplastic	

Notes

Other causes: GB, CIPD, amyloid, paraproteinaemia, acromegaly, sarcoid, and RA.

A Patient with Proximal Myopathy

- General look(according to cause): steroids, thyroid disease
- Ask patient to raise hands above head or combing hair, difficulty standing from sitting.
- Face: steroid use
- Ask about family history

What tests?

- EMG
- X-ray or CT chest for lung
- Vit B12, folate, potassium, and calcium
- Vit D levels

Causes:

- Endocrine: Cushing's, thyrotoxicosis
- Inflammatory: polymyositis/dermatomyositis
- Osteomalacia
- Carcinomatous

Rare causes: McArdle's, mitochondiral myopathy, uraemia

A Patient with Spastic Paraparesis

a. General: bedridden or wheelchair, cerebral palsy if young, cather (bladder involvement)
b. Features of chronic diseases such as motor neuronne disease or multiple sclerosis

   c. Demonstrates upper motor neuron signs: weakness, spasticityand clonus, brisk reflexes with extensor planter or Babinski sign

   d. Fasciculations in MND

   e. Sensory level in cord lesions

   f. Examine back for gibbus deformity, cold absess

Suggested tests:

- Full blood count: anaemia, ESR infection
- Vitamin B12 level, syphilis serology, cancer markers
- Urgent MRI spine

Causes:

- Cord damage(trauma, TB, spinal met)
- Multiple sclerosis
- MND
- Cord tumour
- Friedreich's ataxia
- B12 deficiency
- Parasagital meningioma(no sensory changes)

Specifics:

- Observe gait(if patient can ambulate)
- Find sensory level
- Check for cerebellar signs (Friedreich's ataxia)
- Check sacral sensation
- Examine spine

Notes:

- Other causes: hereditary spastic paraplegia, tropical spastic paraplegia, anterior spinal artery occlusion.

Gait

Common gait abnormalities:

- Spastic paraperesis: patient walks with stiff legs as if cannot bend knees.
- Spastic hemiparesis: affected leg does not move in normal manner, but instead is circumducted at the hip.
- Bilateral foot drop: (due to LMN lesions on both sides) feet lifted high to prevent toes scraping on the floor 'steppage'.
- Cerebellar lesion: patient staggers with a wide-based gait. Heel to toe walking is impossible.
- Parkinsonism: patient walks with stooped gait and 'festinant' shuffling gait. There is no armswing.
- Proximal myopathy: patient walks with waddling gait.
- Posterior column neuropathy: stamping gait

Gait examination:

General	If patient is unsteady, ask the examiner to walk with them and steady them.
<ul><li>Expose with trouser legs above knees</li><li>'Are you able to walk?Do you need a stick/frame?'</li><li>Ask to walk to defined point, turn around, and walk back. Observe type of gait.</li><li>'Can you walk heel to toe?'</li><li>'Can you walk on your toes?' (S1)</li><li>'Can you walk on your heels?' (L5, footdrop)</li><li>Romberg's: 'Would you be able to stand with your eyes closed if I support you?'</li></ul>	If small room, then open door and take patient into corridor. If abnormalities not clear, ask patient to walk fast. Romberg's positive if more unsteady when eyes closed.

Peripheral Nervous System

Lower limbs:

This will either be spot diagnosis such as Paget's, erythema nodosum, pretibial myxedema, etc., or will be neurological. Therefore, if no spot diagnosis obvious, then proceed to full neuro exam.

General • 'Hello, my name is Dr X. I would like to perform some tests on your legs. Do I have your permission to do this?' • 'I would like to see the patient's gait'	Patient normally lying in bed.
Wasting/Inspection • Look specifically for champagne bottle legs, pes cavus, anterior thigh wasting (diabetic amyotrophy) and fasciculations	Small limb from infantile hemiplegia/ childhood polio
Tone • 'Let your legs go floppy and let me move them. Do you have any pain in the hip joint?' • Hands on thigh, and rock leg from left to right. • Hands under knee, and rapidly lift off bed.	Foot should flop when moving leg and not come off bed when lifting leg. If inc tone is this clasp-knife or leadpipe
Power • Hip: 'Lift your leg straight off the bed', 'Stop me from pushing it down', 'Push down against my hands'. • Knees: 'Bend your legs at the knees', 'Pull your heels towards your bottom', 'Push me away'. • Ankles: 'Pull your toes up towards your head and stop me pushing them down', 'Push down against my hands'.	

Reflexes • Knee: Place forearm under both knees and take the weight with them slightly flexed. Gently tap patellar tendon. • Ankle: Fully flex ankle with hand on ball of foot, tap over hand with tendon hammer. • Attempt to elicit clonus at ankles. • Plantar: Stroke firmly up the lateral border of the foot with an orange stick then come round to the big toe.	Clench teeth. Can also elicit ankle jerks by tapping Achilles tendon. Do not use tendom hammer to elicit plantar reflex in exam. If indicated, elicit abdominal reflexes. First movement of big toe is counted as plantar, +ve even if only causes contraction of quads.
Coordination 1. 'Please place the heel of your right foot on your left knee then run it down your shin all the way to ankle and up again.'Repeat on left side.	Do not attempt if gross weakness. Differentiate sensory from cerebellar by repeating with eyes closed.
Sensation • 'Is there any part of your legs that you have noticed in numb or tingling?' • Light touch: 'Please close your eyes and say yes whenever you feel the cotton wool.' Touch at irregular intervals in each dermatome on both legs. • Pinprick: Touch sternum.'Does this feel sharp like a pin?' Touch each successive dermatome.'Does it still feel sharp?'	Map from normal to abnormal. If evidence of cord lesion, then elicit sensory level.

- Vibration: Touch sternum. 'Do you feel this vibrating?' Place on medial aspect big toe, only if patient cannot feel, then test more proximally—medial malleolus ankle, patella, ASIS.
- Joint position: Hold sides of big toe. 'This is up, this is down. Now close your eye and tell me if I move the finger up or down.' Move up and down by small amounts at irregular intervals.

Sensory Level

If suspected spinal cord disease:

1. Use pinprick sensation to determine sensory level (i.e., where sensation becomes normal). If abnormal sensation throughout legs, then proceed to look for sensory level in abdomen/chest.
2. Sensory level on examination often one to twocord segments below actual lesion.

Abdominal Reflexes

- Elicit by lightly stroking the abdominal wall diagonally from the periphery to the umbilicus in each of the four quadrants.
- Upper abdominal reflexes = T9-T10; Lower abdominal reflexes = T11-T12
- Absent in UMN lesions above the level. Also, absent in patients who have had abdominal operations interupting the nerves and normal people.
- Usually lost early in MS and late in MND.

5. Rheumatology Problems at a Glance

Ankylosing Spondylitis

Incidental findings	Iritis, distal arthritis
+ve signs	Question mark posture: loss of lumbar lordosis, fixed kyphosis, compensated extension cervical spine. Protruberant abdomen.
Differential diagnosis	Ankylosing spondylitis
Function	Ability to turn head
Tests	L spine XR – Erosion, sclerosis sacroiliac joints/ syndesmophytes/bamboo spine HLA-B27

Reminder:

- Ask to turn head(whole body turns as block)
- Chest for apical fibrosis
- Heart for aortic regurgitation
- Stand with heels, hips, and occiput against wall. Shober's – marks made 5cm above and below sacral dimples, patient bends forward.

Physical Exam

A physical examination for ankylosing spondylitis often includes the following:

- **Schober's Test:** limited motion in the lumbar spine is symptomatic of AS. The Schober's test measures the degree of lumbar forward flexion, as the patient bends over as though touching their toes. Progressive loss of spinal motion is correlated with X-ray findings.

- **Gaenslen's Test:** sacroiliac pain is often found in the early stage of AS. Gaenslen's maneuver, another name for the Gaenslen's test, stresses the sacroiliac joints. Increased pain during this maneuver could be indicative of joint disease.
- **Chin-Brow Measurement:** this is a method used to measure the spine's curve in the neck. Patient with AS often have necks that angle forward sharply as the spine stiffens. If the doctor is going to use the chin-brow measurement to monitor your angle, the first time he or she takes the measurement will be called your 'baseline'. After that, the doctor will compare each successive chin-brow measurement to the baseline to see if the angle is getting worse.
- **Chest expansion:** when ankylosing spondylitis affects the mid-back region (thoracic spine), normal chest expansion may be compromised. The amount of chest expansion is measured from deep expiration to full inspiration. Measurements significantly less than one inch, which is normal chest expansion, could indicate AS.
- **Range of motion:** to test how well and far your joints allow you to move, the doctor measures the degree to which you can perform movements of flexion, extension, lateral bending, and spinal rotation. Asymmetry may also be noted.
- Other features: achilles tendinitis, plantar fasciitis

Ankylosing spondilitis images: Measurement of flexion deformity when standing against the wall.

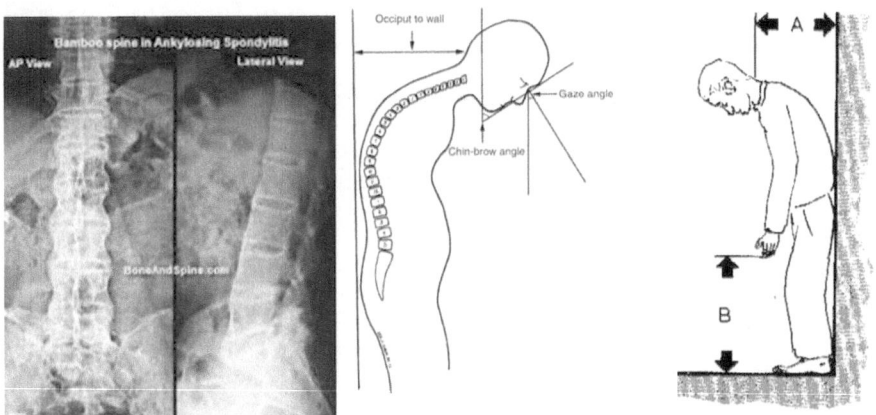

Fig 20: Ankylosing Spondylitis

Schober Test

Technique:

- Patient stands erect with normal posture
- Identify level of posterosuperior iliac spine

    o Mark midline at 5 cm below iliac spine
    o Mark midline at 10 cm above iliac spine

- Patient bends at waist to full forward flexion
- Measure distance between two lines (started 15 cm apart)
- If the distance increases less than 5 cm, then there is an indication that the flexion of the lower back is limited.

The modified Schober method: a technique for assessing spinal motion. Although the technique is reliable (Moll and Wright, 1971), its primary usefulness may be in screening for the very limited mobility that patients exhibit who havediseases like ankylosing spondylitis.

1.) Use a pen to mark the midpoint between the posterior superior iliac spines (PSIS). Then use your tape measure toidentify and mark two points: (1) one that is 10 cm superior to the PSIS, and (2) one that is 5 cm inferior to the PSIS.

2.) As the client flexes the spine as far as possible, measure and record the distance between the superior and inferior marks. If the distance increases less than 5 cm, then there is an indication that the flexion of the lower back is limited.

3.) Similarly, measure and record the distance between the superior and inferior marks as your partner extends thespine as far as possible.

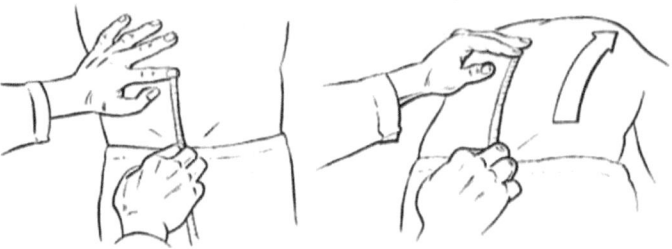

Fig 21: Schber test: ankylosing spondylitis

Psoriasis

Incidental findings	Arthropathy
+ve signs	Nail pitting/oncholysis, salmon pink plaques with silverly scales over extensor surfaces, scalp
Differential diagnosis	• Seborrheic dermatitis • Diaper dermatitis • Onychomycosis • Squamous cell carcinoma • Nummular eczema • Lichen planus • Lichen simplex chronicus • Mycosis fungoides • Subcorneal pustulosis • Pustular eruptions
Tests	CBC, biochemistry, joint X-rays

Notes:

- Subtypes of psoriasis: chronic plaque, guttate, pustular, erythrodermic
- Topical: emollients/keratolytic agents (salicylates)/coal tar/ topical steroids/calcipotriol
- Systemic therapy: UVB, methotrexate, steroids
- Types of arthritis: (i) asymetrical DIP joints, (ii) arthritis mutilans, (iii) asymmetrical oligoarthritis, (iv) ankylosing spondylitis-like, (v) RA-like

Rheumatoid Arthritis

Incidental findings	Cushingoid
+ve signs	'Symmetrical deforming polyarthropathy', swelling MCP, PIP, wasting small muscles, ulnar deviation fingers, boutonniere, swan neck, Z thumb, nail fold infarcts, nodules
Differential diagnosis	RA, psoriasis
Function	Daily household functions, walking
Tests	FBC: anaemia, hypersplenism LFTs: methotrexate s/e RF and anti-CCP antibodies, ANA, HBV, HCV serology, antibodies against citrullinated peptides/ proteins (ACPA). Anti-cyclic citrullinated peptides (CCP)

Notes:

A. Chest for fibrosis, pleural effusion
B. Abdomen for splenomegaly (Felty's, 2° amyloidosis)
C. Peripheral neuropathy, carpal tunnel

Notes:

- Other features: subluxation MCP joint, subluxation ulna at carpal joint, scleritis, episcleritis
- 5 causes anaemia: (i) anaemia chronic disease, (ii) NSAID induced GI bleeding, (iii) bone marrow supression due to methotrexate, (iv) associated pernicious anaemia, (v) Felty's
- Rx: exercise, physio, NSAIDs, DMARDs (methotrexate, sulphasalazine, chloroquine), corticosteroids, anti-TNF
- Indications etanercept/infliximab: ≥ 3 swollen joints and ≥ 3 painful joints and failure of two previous DMARDs (given on own or together)
- Methotrexate side effects: hepatic fibrosis, pneumonitis, cytopenia
- Boutonniere deformity: disruption of central slip of extensor tendon : (b), swan neck (a), ulnar deviation (d)

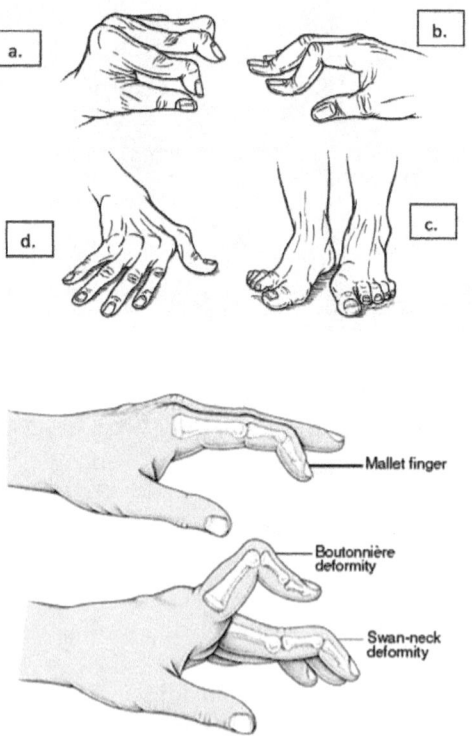

Fig 22: Rheumatoid Arthritis Deformities

Systemic Sclerosis / CREST

Incidental findings	Raynaud's
+ve signs	Tight, shiny skin fingers, sclerodactyly, calcinosis, telangiectasia
Differential diagnosis	Systemic sclerosis/CREST
Function	Hand function
Tests	FBC, CXR Anticentromere, Anti Scl-70

Systemic sclerosis questions:

- Do you have any difficulty with swallowing?
- Do your fingers change colour in the cold?
- Dry eyes/mouth (associated Sjogren's)

Raynaud's questions:

- Do your fingers change colour in the cold?
- What colour do they go?Is there a particular sequence (white/blue/red)?
- How long have you had this problem?
- What is your job?Does it involve vibrating tools?

Notes:

- CREST: calcinosis, raynaud's, esophageal dysmotility, sclerodactyly, and telangiectasia. More favourable prognosis than systemic sclerosis, anti-centromere antibody.
- Limited cutaneous scleroderma = only extremities (10y survival 71 per cent), diffuse cutaneous scleroderma = trunk and extremities (10y survival 21 per cent).
- Rx: symptomatic/immunosupression (cyclophosphamide, methotrexate). No role for steroids.

Systemic Lupus Erythematosis

The classic presentation of a triad of fever, joint pain, and rash in a woman of childbearing age should prompt investigation for SLE. However, patients may present with any of the following types of manifestations[4]:

- Constitutional
- Musculoskeletal
- Dermatologic
- Renal
- Neuropsychiatric
- Pulmonary
- Gastrointestinal
- Cardiac
- Hematologic

SLE

Other types:

- **Discoid lupus erythematosus** is a chronic skin disorder in which a red, raised rash appears on the face, scalp, or elsewhere. The raised areas may become thick and scaly and may cause scarring. The rash may last for days or years and may recur. A small percentage of people with discoid lupus has or develops SLE later.
- **Subacute cutaneous lupus erythematosus** refers to skin lesions that appear on parts of the body exposed to sun. The lesions do not cause scarring.

  It is a distinct lesion that begins as erythematous papules or plaques, and may evolve into papulosquamous lesions resembling psoriasis or annular lesions resembling erythema annulare centrifugum.

- **Drug-induced lupus** is a form of lupus caused by medications. Many different drugs can cause drug-induced lupus. They include some antiseizure, high blood pressure, and thyroid medications; antibiotics and antifungals; and oral contraceptive pills. Symptoms are similar to those of SLE (arthritis, rash, fever, and chest pain), and they typically go away completely when the drug is stopped. The kidneys and brain are rarely involved.

What Would You Look For?

Young patient (female) complains of fatigue, joint pains, and skin rash.

Face:

1. Butterfly (malar) rash, mouth ulcers, pallor
2. Scarring alopecia, fundus: vasculitis
3. Eyes: Sjogren's syndrome (dry eyes). Fundus: retinal haemorrhages and cytoid bodies(white exudates)

CNS: Chorea, fits, psychosis, meningitis, focal lesions

Respiratory: pneumonitis, fibrosis, pleurisy andpleural effusion, pulmonary oedema.

Heart: lookfor pericardial friction rub, murmur, Libmansacks endocarditis. Also, cardiac enablement and signs of heart failure

Renal: look for signs of renal failure

Abdomen: hepatomegaly

Muscles: look for proximal weakness

Legs: pitting oedema (nephritic syndrome)

Fingers: Raynaud's, infection and gangrene (vasculitis), arthritis

Miscellanous Cases: Station 5

Thyroid Diseases

How to perform during examinations?

o   First: Approach

Before entering, read the instructions carefully, and make an intial plan.

1.   Enter the room and wash hands
2.   Greet (you may shake hands if situations allows)
3.   Introduce youeself
4.   Explain the purpose and get permission

o   General examination:

•   Comment on the appearance of the patient: thin, wasted, anxious, face shows exophthamus(hyperthyroidism), overweight, dull apathetic facial look, coarse features, thin hair (hypothyroidism).

o   Examine for eye signs:

•   Observe exophthalmus, see from behind and front
•   Lid retraction: you can see sclera above the cornea (normally covered by the upper eyelid). This is due to sympathic stimulation of the muscle levator palpetrae.
•   Examine for lid lag: stand at about half a metre from the patient and ask the patient to follow your finger, moving from upwards, downwards at a reasonable rate. In thyrotoxicosis, the upper lid lags behind the descending eyeball.

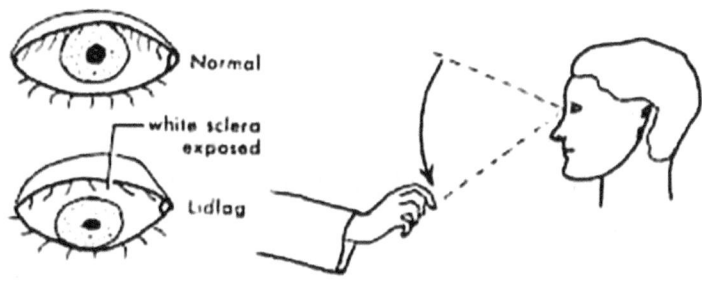

Fig 23: Lid Lag

Grave's Ophthalmopathy (exophthalmos, periorbitaloedema, conjunctival injection, chemosis)

- When you you finish examiningthe face, examine the hands: warm and sweaty, fine tremor, palmar erythema, and clubbing (thyroid acropathy). Elicit fine tremor.

Testing for Fine Tremor

Fine tremor is a feature of hyperthyroidism. Ask the patient to outstretch his hands with dorsum of the hand facing upwards. Keep this position for awhile.

If the tremor did not appear, put a piece of thin paper over the hands, and observe the shaking of the paper. Also, you can place your palms on the dorsum of the patient's hands and you may feel the tremor.

The pulse: usually, there is tachycardia, collapsing pulse, or atrial fibrillation.

Then look at the legs for the waxy indurated swelling: pretibial myxedema (thyroid dermapothy).

Signs of Hypothyroidism

Look at the face:

Thickened and coarse facial features with puffy eyes. One old sign is St Ann's sign. This is the loss of hair from outer third of eyebrows, but this is not a reliable sign.

- The patient may have xanthelasma due to hyperlipidemia.
- Feel the pulse. Usually, there is bradycardia.
- Patients with myxedema may have carpal tunnel syndrome. So you may test for that using Tinel's or Phalen's test.

Inspect the neck for goiter (nodular or symmetrical), scars of previous surgery(hemi/total thyroidectomy). If there is goiter, detect any nodularity or nodules.

Legs: non-pitting oedema. Also, slow relaxation phase of the ankle jerk. This is best demonstrated with the patient kneeling on a chair or bed with the feet hanging over the edge, and the examiner standing behind the patient.

Examination of the Thyroid Gland

Inspection

Sitthe patient comfortably in chair, give him a glass of water, and tell him totake a sip, keep in his/her mouth, and swallow when instructed. Observe the upward movement of the goiter. The thyroid moves upwards on swallowing since it is enveloped in the pretrachealfascia, which is attached to the cricoid cartilage.

A thyroglossal cyst will move upwards on swallowingand protrusion of the tongue and can be transilluminated. Then palpate the neck for lymph nodes.

Palpation

- Ask the patient forpermission to feel the neck. Be getle and apologise if the patient feel a discomfort.
- Examinethe patient from behind with the neck slighltly flexed.
- Identify the isthmus of the thyroid with your right index and middle finger. It lies about two fingers width below the thyroid cartilage.
- Attempt to gently paplate each lobe at a time. It is better to palpate while the patient swallows water.
- Extend palpation upwards along the medial edge of the ternomastoid muscle oneither side to look for a pyramidal lobe.
- Describe any thyroid swelling completely: size, consistency, and tenderness.
- Percussover the upper sternum to assess any retrosternal extension of goitre.
- If there is a large goitre with retrosternal extension, test for evidence of compression by eliciting Pemberton's sign. This is done by asking the patient to raise his arms above the head. It is positive if there is facial congestion, syncope, or dizziness.

If there is a thyroid swelling, auscultate for presence of a bruit. This is a vascular systolic murmur, which is seen in Grave's disease. Verify this by auscultating over the aortic area to make sure that this bruit is not a conducted murmur of aortic stenosis.

Complete the examination of the neck byfeeling for lymph nodes in all areas. If enlarged, may indicate thyroid cancer or acute thyroiditis.

How to do it? From behind, palpate for the supracalivicular, deep cervical. Then submental, submandibular, and preauricular. Then approach the patient from front and palpate for the posterior triangle, posterior auricular, and suboccipital lymph nodes.

Assess the size, consistency (soft or rubbery as in Hodgkin's lymphoma or hard in malignanacy), discrete (reactive, lymphoma, or infectious mononucleosis), or matted (in TBand malignanacy). Also, observe any sinuses or tenderness. Decide if the lymph nodes are mobile or fixed to the skin and deep tissue (malignancy). Finding any enlarged nodes will prompt you to examine for axillary and inguinal lymph nodes. Genaralised lymphadenopathy is associated with TB or lymphoma.

How to Examine for Axillary Lymph Nodes

Examine the right axilla from the right side of the patient and vice versa. Askthe patient to relax his right arm on your examining hand (left hand) a little away from thechest wall (use your left hand to examine the right axilla and vice versa). Gently place yourfingertips into the vault of the axilla, and then draw them downwards feeling the **m**edial, **a**nterior, and **p**osterior walls in turn

Observe if there is a scar caused by biopsy and for radiotherapy marking. If there is generalised lymphadenopth, ask permission to examine the abdomen for liver and spleen enlargement. Inexams, ask permission to examine the chest and request X-ray to look for paratracheal and mediastinal lymphadenopathy.

Additional:

- Examine for tracheal deviation if there is neck swelling.
- If you find a thyroidectomy scar, test for hypocalcemia caused by hypoparathyroidism using Chvostek's test. Tap the facial nerve about three to five centimetre from the ear. It is positive if there is twitching of the lips and facial muscles.

Futher Assessment of Thyroid Disorders

- Tendon jerks (ankle): brisk in hyperthyroidism and slow elaxation phase in hypothyroidism. This is best demonstrated with the patient kneeling on a chair or bed.
- Test forproximal myopathy(thyrotoxicosis). Ask the patient to raise arms up like chicken wings and try to overcome this.
- Eye examination: test for lid lag and lid retraction. Also any congestion.
- Test eye muscles for ophthalmoplegia.
- Visual field forbitemporal defects indicationg pituitart tumour.
- Examine the fundus: optic atrophy
- Chech blood pressure: high in both hyperthyroidism and hypothyroidism
- Check urine for sugar. DM and thyroid diseases are both autoimmune.

Miscellanous Short Cases and Examples for PACES

Station 5: BBC and Comprehensive Patient Evaluation in Other Boards

o Anaemia

A 64-year-old lady complained oftiredness and shortness of breath. Hb was 6.7 G.

Asked to take focused history and examination:

- After doing the basics, I asked about symptoms symptoms of blood loss.
- Expected to ask also about complainsof haemolytic anemia, blood diseases, and pernicious anaemia.
- Lao, I would ask about past history of blood transfusion, gastric surgery, and medications(NSAIDs). Enquire about family history of anaemia or blood diseases.

Examination:

- Examinethe face for pallor, jaundice(haemolysis), telanjectasia, tongue appearance, and vitamin deficiency (angular stomatitis).
- Examine hands for pallor, nail changes.
- Then complete the history. Patient mentioned in the FH that her dad had smiliar problems and suffered from nosebleed, which she also had over the previous few weeks.
- The patient has obvioustelengiectasia on her lips. Also same in mouth and andbuccal mucosa.
- Examinethe neck for lymph nodes andthe abdomen for hepatosplenomegaly(blood and live diaeases).
- Attempt to examine the chest for *A–V* malformation (cyanosis).
- Explain to the patient the likely diagnosis (hereditary haemorrhagic telangectasia), and what investigations are needed.
- The patient might ask of any useful treatment. You have to tell the patient that there is no cure since it is inherited, so she can receive blood transfusion and iron supplements. If significant GI bleeding occurs, endoscopy can be done with possible endoscopic treatment.

o Patient with loss of vision

72-year-old gentleman referred by GP because he has deteriorating vision.

How to proceed?

- Take focused history.
- Ask about onset, pattern, any eye pain, headache, or neurological symptoms.
- Examine the eye for redness, squint, or bulging.
- Do proper full fundus examination.

- This gent had near complete loss of vision bilaterally, positive family history.
- Fundus showed retinitis pigmentosa.
- Give a plan of management: home help (occupation therapist)

o Jaundice in a young man

- Take focused history: onset, duration, progress, itching, abdominal pain, and family history.
- Consider liver diseases and blood diseases (haemolytic anaemias).
- Examine the face: eyes for extent of jaundice, Kayser-Fleischerring, xanthelasma, and pallor. Examine the neck for lymph nodesand abdomen for enlarged organs, ascites, and dilated veins.
- Patient reported off and on jaundice since childhood. Has family history of jaundice.
- Discuss differential diagnosis: most likely haemolysis, spherocytosis (patient has splenomegaly). Others: Thalassemia, leukemias, and liver diseases.
- Discuss the management plan. Invistigations: CBC, peripheral blood film, HB electrophoresis, andpossible bone marrow. Management: supportive and possible splenectomy.

o Station 5 PACES (Comprehensive patient evaluation other boards)

You are asked to evaluate this 54-year-old man with rheumatoid arthritiswho presents with increasing shortness of breath.

- Do the basics: GINEP(greet, introduce self, confirm name of patient, permission).
- Take focused history about onset, progression, and grade of shortness of breath. Ask about associated cardiac and respiratory symptoms: chest pain, palpitations, cough, wheezes, and paroxysmal nocturnal dyspnea.

THE ART OF PHYSICAL EXAMINATION

- Elicit past history. First, ask about details of his RA. Is it controlled? What medications? Then any cardiac, respiratory, and other chronic illnesses.
- Ask about medications for RA and others.
- Examination: Observe face for steroid effect, cyanosis, pallor, and mouth for vasculitis, small bucurred mouth of systemic sclerosis.
- Examine hands and other joints for RA deformity, systemic sclerosis (skin), nail changes, and clubbing.
- Look for signs of heart failure, JVP, basal rales, and auscultation of the heart.
- Examine lower limbs and sacrum for oedema.
- Explain to the patient the likely diagnosis and the chances of treatment(i.e., management plan).

Diagnosis: The patient has pulmonary fibrosis and evidence of right-sided heart failure (cor pulmonale). Pulmonary fibrosis is likely to be caused by RA (or drugs for RA).

o More examples for PACES Station 5: BBC and comprehensive patient evaluation in other boards.

This patient complained of joint pains and skin rash.

- Follow the same style as above example.
- Basics: GINEP
- Ask about details of joint pains, pattern, severity, any symptoms of inflammation, fever, and effect on daily activities.
- Get the details of the skin rash, distribution, type (popular, macular, or vesicular), itching.
- Past medical history
- Examine the face for pallor, effect of steroids, mouth for vasculitis.
- Examine the joints for signs of inflammation and deformity.
- Examine the hands for deformity and nails for pitting.

- Examine the skin to identify the type of rash: distribution, colour, scaly, or itchy.
- Explain to the patient the likely diagnosis and the management plan.

Diagnosis: Posoriasis with arthritis.

o More examples for PACES Station 5: BBC and comprehensive patient evaluation in other boards.

The following is a frequently seenscenario in exams.

A 49-year-old lady, known IDDM, experienced some visual problems in her left eye. She is very concerned about this problem.

- You are the physician in the diabetes clinic. You are required to take focused history and examination and to explain the problem to the patient.
- Gather information. Ask about duration and treatment of DM. Enquire if the patient is compliant with treatment (insulin).
- Get the details of the visual problem. Is it loss of or blurred vision? Any scotomata, floaters, double vision, or painful eye?
- Ask about symptoms in the CNS (weakness, pins and needles, numbness) in the gastrointestinal system (indigestion, fullness, diarrhea, or constipation), genitor-urinary system, and cardiovascular symptoms (chest pain, palipitations, shortness of breath). Also ask about intermittent claudication.
- Focused examination: general look, face for pallor cyanosis, site of insulin injection. Quick CNS for neuropathy.
- Full eye examination for eye movement, field of vision, and pupil. Fundus examination for retinopathy, maculopathy, and disc changes.

271

- Infact, the patientstates that she is on insulin pump, and her latest HbA1C is 7.2. She reports that she had a prvious myocardial infarction and a minor stroke.
- Examination showed that she has loss of hypoglycemia awareness. Sensorimotor diabetic neuropathy is detected, but no foot ulcers.
- Eye: reduced visual acuity in the left eye, has red reflex. Bilateral laser scars and maculopathy in the left eye.
- Explain the findings and available treatments to the patient.
- Deal with concerns of the patient. She asks if she is going to have complete blindness, and if the treatment will reverse these changes.

7. More examples for PACES Station 5: BBC and comprehensive patient evaluation in other boards.

This 53-year-old gentleman, known HIV, complained of vision problem in his right eye.

How to proceed?

- As mentioned in the previous example, gather focused history about the background of the patient (HIV) and its treatment.
- Get details of the visual problem: onset, visual acuity, floaters, colour blindness, etc.
- Examine face for pallor and fundus for retinopathy, cytomegalovirus.
- The patient reports frequent headaches associated with the start of visual problem.
- Examination of the eye revealed bilateral papilloedema and retinopathy. Cytomegalovirus.
- Suggested investigations: blood workup (CD4) and possible CT or MRI of brain.
- Suggest lines of management.

8. More examples for PACES Station 5: BBC and comprehensive patient evaluation in other boards.

A 42-year-old male patient referred because of frequent headaches and a BP of 185/125. You are required to take focused history and examination.

- Do the basics and take focused history about details of the headache and associated symptoms. Ask about possible causes of secondary hypertension.
- Examination: features of acromegaly
- Test visual fields (bitemporal hemianopia)
- Look for scars of previous surgey for acromegaly (transphenoidal resection).
- Examine for carpal tunnel syndrome.
- Discuss how you treat his hypertension.

o **Quick brief examples for PACES Station 5: BBC and comprehensive patient evaluation in other boards.**

- Young lady with skin lesins. Sha has chronic diarrhea and weight loss. **Dianosis: Pyoderma gangrenosum associated with ulcerative colitis.**
- Pregnant lady presents with TIA, previous two abortions, and DVT twice.

Diagnosis: Antiphospholipid syndrome. Examine eye for pallor, mouth for ulcers, hand for clubbing, and abdomen for scars of surgery, masses. Discuss investigations (colonoscopy) treatment during pregnancy.

Lady with tiredness and HB 8.6 G. **Ask about:**

- Family historyof similar problem?
- GI bleeding
- Epistaxis

- Headaches, stroke (cerebral AV malformation)
- Focused examination, looking for: mouth, telengiectasia, tongue for pallor
- Examine chest for bruits
- Cardiovascular examination for high output cardiac failure

Diagnosis: Hereditary haemorrhagic telangiectasia

Discussion:

- Complications: epistaxis, GI haemorrhage, iron deficiency anaemia, haempoptysis, subarachnoid, high output cardiac failure (due to hepatic artery-vein shunt)
- Genes: endoglin-β, activin receptor-like kinaes 1 (ALK-1). Both expressed on vascular endothelial cells and implicated in signalling to members of TGF-beta family.

Marfan's Syndrome

Patient with poor vision

Inspection:

- Tall patient with archnodactyly wearing thick glasses (myopia)
- Blue sclera, enophthalmos, and dislocated lens
- High-arched palate
- Pectus excavatum or carinatum
- Skeletal deformity like scoliosis
- Skin: stretch marks
- Examine the cardiovascular system for aoric regurgitation, mitral valve prolapse, and evidence of aoric dissection.
- Chest: pneumothorax

What differential diagnosis?

- Homocystinuria (less lens dislocation, mental retardation)
- Ehlers-Danlos syndrome
- Multiple endocrine neoplasiatype B2

Special tests:

1. Thumb sign(Steinberg sign), clenched thumb in fist. A positive thumb sign is where part of the thumb is visible beyond the ulnarborder of the hand, caused by a combination of hypermobility of the thumb, as well as a thumb which is longer than usual
2. Wrist sign (Walker Murdoch sign): the thumb and little finger overlap when grasping the wrist of the othe hand. A positive wrist sign is where the little finger and the thumb overlap, caused by a combination of thin wrists and long fingers.

Other investigations:

- FBC, CXR
- Anticentromere,
- Anti Sel-70
- Echocardiogram: need annual echocardiogram. Prophylactic replacement of aortic root when diabeter 50-55mm

Management:

- Need annual echocardiogram: Prophylactic replacement of aortic root when diabeter 50–55mm
- Beta blockade to slow rate of aortic root dilation
- Mutation of fibrillin (FBN1) gene on chromosome 15

This patient has high blood pressure and poor hearing: Neurofibromatosis

- Examine skin: multiple cafe au alait spots
- Lish nodules
- Axillary freckles
- Kyphoscoliosis

What else would you do?

- Test for visual acuity, fundoscopy for optic nerve glioma
- Hearing for acoustic neuroma
- Would like to know blood pressure (pheochromcytoma)

What tests?

CBC, kidney function,
US abdomen (polcystic kidney)
CT/MRI brain(aneurysm)

Notes

- Complications: malignant transformation, hypertension, spinal cord/nerve root compression
- Screening: Lisch nodules in 100 per cent of patients >twenty years
- Type two usually bilateral acoustic neuroma or unilateral acoustic with menigioma/neurofibroma/schwannoma
- Other features: lung cysts, retinal hamartomas, skeletal abnormalities, intellectual disability

Paget's Disease

Quick takes:

- Examine this who complained of deafness
- Observe the facial look: shape of the head (large triangular)
- Hearing aid
- Skeletal: kyphosis
- Legs: Cabre tibia, feel for warmth
- Examine heart for aoric regurgutation and high output heart failure
- Fundus: optic atrophy or angiod streaks

What test to suggest?

- Urine analysis: renal stones
- Plain X-rays
- Bone scan
- Calcium, Alkaline phosphatase
- Old photos

What Complications?

- Fractures
- Cord compression/root lesions
- Cranial nerve signs
- Hypercalcaemia
- High output cardiac failure, sarcomatous change
- Hearing loss can be either involvement of ossicles or compression of VIII by skull foramina
- Treatment: bisphosphonaes, calcitonin

Other Problems

Pseudoxanthoma Elasticum

Patient may present with GI bleeding, abnormal joint movements, stroke, visual problems, MI, etc.

Main signs:

Thick kerainised and creased neck (chicken skin)

What test to confirm daignosis?

Skin punch biopsyusingVan Gieson stain shows fragmentation elastic tissue

Further points:

- Examine peripheral pulses
- Fundus for angioid streaks (also in Paget's, sickle cell, Ehlers-Danlos)

Notes

- Remember to look at neck in patient with generalised pigmentation in retina.
- Complications: upper GI haemorrhage, visual loss, MI, stroke, claudication, miral valve prolapse, renovascular hypertension.

Take home message:

Lastly, I remind you of being professional, meticulous, and curteous when you communicate with or examine patients. Prepare well for your exams and read these pearls with interest, and try to apply what is mentioned. See more patients as a group and practise communication, history, and history taking.

Good luck to all our students and doctors on training.

Professor Mohamed-Elbagir Khalafalla Ahmed

MBBS(University of Khartoum, Sudan), MD(University of Khartoum, Sudan)

MRCP(UK), FRCP(UK), FACP(USA)

Email: mohamedelbagir@live.com

January 2017

www.ingramcontent.com/pod-product-compliance
Lightning Source LLC
Chambersburg PA
CBHW020732180526
45163CB00001B/203